Rodrigues Ottolengui

Methods of Filling Teeth

An Exposition of Practical Methods

Rodrigues Ottolengui

Methods of Filling Teeth
An Exposition of Practical Methods

ISBN/EAN: 9783337338077

Printed in Europe, USA, Canada, Australia, Japan

Cover: Foto ©berggeist007 / pixelio.de

More available books at **www.hansebooks.com**

An Exposition of Practical Methods which will Enable the Student and Practitioner of Dentistry Successfully to Prepare and Fill all Cavities in Human Teeth.

BY

RODRIGUES OTTOLENGUI, M.D.S.

WITH TWO HUNDRED AND THIRTY-SIX ILLUSTRATIONS

Giving Exact Representations of all Classes of Cavities and their Management.

PHILADELPHIA:
THE S. S. WHITE DENTAL MFG. CO.
LONDON:
CLAUDIUS ASH & SONS, LIMITED.
1892.

Copyright, 1891, by R. OTTOLENGUI.

Entered at Stationers' Hall, London.

PREFACE.

THERE are already so many text-books that the question might be asked, "Why another?" My reply gives my excuse for my intrusion. Without designing to criticise the methods of other writers, I would yet call attention to the fact that many have given us works which are largely compilations. These authors have seemed loath to leave anything unsaid which is pertinent to their subjects. In their efforts to be fully comprehensive they have quoted freely from others, giving pros and cons by men of equal authority, till the student who is a beginner is bewildered in his effort to choose. To avoid this, I decided to describe in my book only such methods as I have myself tested, believing that the student will be more benefited by adopting a single successful mode of practice than by essaying the various methods of many men.

This has involved a two-fold result. First, and most important, the teaching becomes dogmatic.

The charge has been freely made that "writers are not skillful dentists." This is because theory and practice are so often at variance. I have endeavored to write a work which would be as practical as words could make it. There is not a case described that has not occurred in my practice. There is not a method advocated that I have not tested.

The second result is that I do not give detailed directions for carrying out methods which I have not attempted. This of course makes the book incomplete from that standpoint; but I prefer this to being quoted as authority for that which I have not myself tested, as too many have been already. As an example of such omission, it will be observed that I do not describe methods of using non-cohesive gold foil. I can only say in defense that I have never used non-cohesive foil, and let that excuse my not treating of it. I will reiterate, however, what I say in the body of the book, that I have never seen any need of it, nor found any man who could prove its necessity. I do not think my patients have suffered because of my lack of knowledge in this direction.

Because of the fact that my work first appeared in serial in the *Dental Cosmos*, I am enabled here to reply to one or two criticisms which have been printed in society reports. One gentleman quotes me as advocating a broad contact-point in approximal fillings. In this, if he was correctly reported, he has misrepresented me. My advocacy of such a contact is in connection with a specified condition only, and the position which I take is one which I am ready to defend clinically or otherwise at any time. This would be an inappropriate place to discuss it.

Another gentleman is reported to have said that I advise students to have as few instruments as possible, and that I myself fill teeth with a broken instrument. The first statement is accurate. I think that a man beginning the practice of dentistry should not purchase many instruments until he has had the experience which will lead him toward a wise choice of such an assortment as will best suit his individual peculiarities and requirements. The second statement is inaccurate. I simply say that a good filling may be inserted with a broken instrument. The point here was in reference to whether the point of a plugger should be serrated or smooth. A broken point is neither the one nor the other, yet may be a good point. While it is not my practice to fill teeth with a broken instrument, as suggested, I could easily demonstrate that as good a filling can be inserted in that way as with the best new plugger. Again I say it is the man, not the tool.

Let me say here, as I have said in the body of this work, I do not make any broad claims for originality in connection with the methods described. If there is any originality at all, it is in the method of teaching, rather than in the thing taught. To offer the profession, and especially those just entering it, a work advocating methods entirely my own would be to ask the adoption of modes of practice not in common use, and therefore not sufficiently well tested.

I have no sympathy with those who are constantly crying out, "That is my method; I invented it." The chief interest to the student must always be in a thorough knowledge of the method itself, rather than in the name of its originator.

To those, therefore, who will find a description of their original methods in my book I have only to say, "Gentlemen, I thank you for what you have taught to me, and am now in turn trying to teach it to others." I wish also to extend my heartiest gratitude to the hundred or more dentists of this country and Europe who, though strangers to me, have written me kindly letters of approval during the progress of this work. Many times have such letters proved a solace to me when hand and brain were tired, and so have been an incentive to continue. I thank them.

RODRIGUES OTTOLENGUI, M.D.S.

115 MADISON AVENUE, NEW YORK, September 1, 1892.

CONTENTS.

CHAPTER I.

GENERAL PRINCIPLES INVOLVED IN THE PREPARATION OF CAVITIES—REMOVAL OF DECAY—RETENTIVE SHAPING—INTENTIONAL EXTENSION—DIRECTIONS FOR FORMING CAVITY-BORDERS—SHAPING ENAMEL MARGINS.. 1

CHAPTER II.

GENERAL PRINCIPLES INVOLVED IN THE FILLING OF TEETH—METHODS OF KEEPING CAVITIES DRY—THE RUBBER-DAM—LIGATURES—CLAMPS—LEAKAGE—THE NAPKIN—CHLORO-PERCHA—WEDGES *vs.* SEPARATORS—THE USES AND DANGERS OF MATRICES................................. 27

CHAPTER III.

THE USES OF VARIOUS FILLING-MATERIALS—METHODS OF MANIPULATION—MATERIALS OF MINOR VALUE—GUTTA-PERCHA—OXYCHLORIDE OF ZINC—OXYPHOSPHATE OF ZINC—AMALGAMS—COPPER AMALGAM—GOLD... 47

CHAPTER IV.

THE RELATIVE VALUES OF CONTOUR, AND FLAT OR FLUSH FILLINGS—THE V-SHAPED SPACE IN ITS RELATION TO THE GINGIVA—THE RESTORATION OF SUPERIOR LATERAL INCISORS—SLIGHT CONTOURS—REGULATION OF TEETH BY CONTOUR FILLINGS—DEPARTURE FROM ORIGINAL FORM—TRUE CONTOURING—TREATMENT OF MASTICATING SURFACES—CONTOURING WITH GOLD—WITH AMALGAM—WITH THE PLASTICS IN CONNECTION WITH GOLD PLATE—USE OF SCREWS—CASES FROM PRACTICE REQUIRING ODD METHODS............................. 83

CHAPTER V.

SPECIAL PRINCIPLES INVOLVED IN THE PREPARATION OF CAVITIES, AND IN THE INSERTION OF FILLINGS—CONSIDERATION OF APPROXIMAL CAVITIES IN INCISORS—IN CUSPIDS—IN BICUSPIDS—IN MOLARS. 104

CHAPTER VI.

SPECIAL PRINCIPLES INVOLVED IN THE PREPARATION OF CAVITIES AND THE INSERTION OF FILLINGS—CAVITIES IN THE MASTICATING SURFACES—INCISORS—TREATMENT OF IMPERFECTIONS—OF FRACTURES—OF ABRASIONS—OF MALFORMATIONS—CUSPIDS—BICUSPIDS—MOLARS—OXYPHOSPHATES IN COMBINATION WITH GOLD—UNITING TEETH BY BAR AND FILLING.. 126

CHAPTER VII.

SPECIAL PRINCIPLES INVOLVED IN THE PREPARATION OF CAVITIES AND THE INSERTION OF FILLINGS—SENSITIVENESS AT THE TOOTH-NECK—EROSION—GREEN-STAIN—TRUE CARIES—FESTOON CAVITIES—THE LABIAL SURFACE—THE PALATAL—THE LINGUAL—BUCCAL CAVITIES—TEMPORARY FILLINGS—THE FINISHING OF FILLINGS....... 151

CHAPTER VIII.

METHODS OF FILLING THE CANALS OF PULPLESS TEETH—A STUDY OF TOOTH-ROOTS—METHODS OF GAINING ACCESS TO AND PREPARING CANALS—WHEN AND HOW TO FILL ROOT-CANALS........................... 176

INDEX.. 197

METHODS OF FILLING TEETH.

CHAPTER I.

GENERAL PRINCIPLES INVOLVED IN THE PREPARATION OF CAVITIES—REMOVAL OF DECAY—RETENTIVE SHAPING—INTENTIONAL EXTENSION—DIRECTIONS FOR FORMING CAVITY-BORDERS—SHAPING ENAMEL MARGINS.

Now that the dentist is no longer to be denominated the "knight of the forceps," it is fundamentally essential that he who would become a conscientious practitioner should be able to determine whether a tooth be salvable by the insertion of a filling ; to decide which of the many materials now in use will best attain the end in view ; to properly prepare the cavity for the reception and continued retention of the filling, and be capable of scientifically and skillfully placing it so that it will be as nearly perfect as the attending conditions will permit. This much has been accomplished by a goodly number of the dentists of the past and present. Something more will be expected of the dentist of the future. He will be asked to abandon the assertion, "Madam, your tooth has decayed around my filling, *but the filling was all right."*

Undoubtedly there are teeth in which it may be impossible to prevent recurrence of decay, but it is equally true that in too many cases when the "tooth decays around the filling" the filling was *not* "all right." It is opportune, therefore, to discuss these questions more in detail than has been done heretofore.

Those whose fillings are frequently returned to them in a leaky condition are compelled to adopt one of two propositions : Either their work is inefficient, or else the teeth upon which they have operated are of poor quality. It is but human to lean toward the latter explanation. The position, however, is rarely tenable. The argument used is this : "If the tooth decayed when it was perfect, why should it not do so after it has been filled? I cannot be expected to build better than did the Creator." This sentiment was loudly applauded at a national meeting, leaving the impression upon the mind of him who analyzed the situation that there were many present anxious to adopt this specious excuse for the failures which had attended their efforts. The fallacy lies in this: While it is perhaps true that no material exerts any therapeutic influence beyond the mere mechanical stopping of a hole and restoration of contour, it is also true that, given a tooth, and certain conditions under which it is attacked by caries, the caries will occur invariably at a specified situation. Therefore, when the cavity is filled scientifically the tooth is safer than ever, because the vulnerable point is now occupied by a material which will resist destruction by caries. If decay occurs along margins, it is because those margins were improperly made either as to shape or position, or else because the filling was unskillfully inserted or finished. An ideally completed filling is one which is given as high a polish as the material used will permit. Those who argue for "dull finish," because less conspicuous, forget that "high polish" means smoothness, which quality is a prerequisite.

The student watching his preceptor is almost invariably impressed with the idea that only a few principles are involved, and that the operation of filling a tooth is purely mechanical. As soon as he acquires the knack of packing gold and producing a polished surface afterward, he considers that "he knows it all." It is only after several years of bitter experience at refilling teeth for his own patients, that he begins to suspect that perhaps there is more in this branch of dentistry than his mind had grasped. Let us now consider the subject in detail, from the point of view that there is more involved in it than mere mechanics.

GENERAL PRINCIPLES INVOLVED IN THE PREPARATION OF CAVITIES.

Removal of Decay.—When a cavity has been properly prepared, the tooth is half filled. The most beautifully polished, solid, well-formed filling will fail if the cavity has not been skillfully shaped.

To the mind of the layman it would seem idle to discuss the propriety of leaving any decay in a cavity. It is not uncommon to have a mother say, "Doctor, please be sure to take all the decay out, as I don't want Willie's tooth to trouble him again." To her mind, safety lies in thorough cleansing. It seems a rational proposition, yet it has been argued by high authorities that there are frequently occurring cases where it is best not to remove all the decayed dentine.

This is a grave error. With rare exceptions it is imperative that every trace of caries should be obliterated. The tooth about to receive a filling should be as wholly healthy as it can be made.

It has been claimed that decay covering a pulp may be left in place and sterilized with safety and advantage. This sterilizing is usually done at the sitting at which the filling is placed. I have followed this advice in a few cases, where, in the front of the mouth, it seemed best to take every precaution to avoid destruction of the pulps and consequent discoloration. In every case I have afterward removed the fillings, because of a bluish appearance which subsequently presented, showing that despite the fact that the edges were yet perfect, decay was progressing internally. This shows that a perfect gold filling will not stop decay if carious dentine is left in the cavity. Recent reports from Professor Miller are in harmony with this experience, since he shows that the germicides upon which most reliance has been placed are ineffectual unless left in a cavity much longer and in greater quantity than has been our practice. When our chemists shall have discovered for us a sterilizing agent the use of which will assure us of a discontinuance of carious action, *in already carious dentine*, then, and not till then, will there be any argument worth listening to against the assertion that *it is malpractice to fill over decay*.

There is, however, a delicate distinction to be made between dentine which is carious, and that which may be slightly affected because of its juxtaposition with caries. As fast as caries advances, the dentine suffers alteration. This action is physiological, and is nature's method of defense against the inroads of the enemy. A retrogressive metamorphosis occurs. The lime-salts are partly dissolved out, the dentine returning toward an embryonal condition. If the progress of the disease be slow, a redeposition of lime-salts occurs, the new tubuli being built up of a form better able to resist caries. Where this is successful, we observe that condition which we term "arrested decay." The surface is brownish in color, and flinty; the tooth is safe; nature has cured the disease. We may therefore cut down to a stratum of dentine which is undergoing this change, which is easily cut, but not, strictly speaking, decayed. It is highly important in these cases not to excavate deeply enough to expose the pulp. The operator must have the knowledge to enable him to determine when to stop. The experienced practitioner recognizes the condition by sight and touch, but a definite description may serve to guide others.

The softening of non-carious dentine occurs most frequently in those teeth from which we remove carious dentine in leathery layers. To excavate such a cavity, use spoon excavators only. Never employ a hatchet, or a hoe, or an engine-bur, unless needed for shaping after the decay has been removed. With a spoon begin at a point farthest away from the pulp, and gently lift the outer edge of a layer. Having thus disengaged it, proceed to lift it around its whole circumference, and then work gradually toward the center till it can be taken out of the cavity. Repeat this as often as a layer can be started at its circumference. As soon as the last distinct layer has been removed, scrape all the walls vigorously, removing even the softened dentine which may be clinging to them. The cavity will now be clean, but the bottom of it will be soft. Still using the spoon, scrape the bottom very gently, starting at the circumference and approaching the center, removing all small particles which may be thus disengaged, without actually cutting. A reliable germicide should now be used on cotton, the cavity sealed with a phosphate cement, and so left for two or three days, when the filling may be inserted with safety.

Non-carious dentine may be found beneath other than leathery decay. If the carious portion is brown, or black, the demarcation is discoverable by the color, softened dentine being usually of nearly normal color. When it is found underlying decalcified enamel, the enamel will come away as a white, chalky powder, and the dentine beneath need only be removed as required for the retention of the filling. This condition is rarely observed except under "green-stain," and the destruction is a decalcification, rather than true caries. A corundum stone should be used to remove the stain, when the decalcified enamel will readily be distinguished by its white color. The dentine beneath is non-carious, but as the destruction is superficial there is little risk of exposing the pulp, for which reason no special caution is needed beyond the usual care when operating on healthy dentine, not to cut deeper than is actually necessary to correctly shape the cavity. It may be said in passing that this decalcification of enamel due to or accompanying "green-stain" is usually associated with highly sensitive dentine. This is fair presumptive evidence that the dentine is undergoing a change which by loss of lime-salts leaves a disproportionately large amount of soft tissue. In this particular class of cases the sensitiveness may usually be controlled sufficiently by hot-air blasts to allow rapid preparation of the cavity with sharp engine-burs.

Retentive shaping.—The cavity cleansed of decay, the next important object is to so shape it that the filling cannot be dislodged mechanically after it has been inserted. To accomplish this sometimes taxes the utmost ingenuity of the most experienced, so that binding rules cannot be formulated to cover all conditions. It will therefore be best, in order to describe methods covering a wide field, to take up individually the more common cavities, but before proceeding to that discussion I shall present here a few general principles.

The great desideratum is to so form the cavity that the visible external surface of the filling, when placed, shall have a smaller diameter than some portion which is within the cavity. We should thus have a mass occupying a cavity whose orifice would not permit the passage of its greatest diameter. Such a filling could not be removed mechanically, except in pieces. If the material therefore were durable, the filling would be permanent as long as the opening was not enlarged by decay or fracture. There are, however, other considerations which may make it imprudent, or impossible, to follow this rule, as, for example, when such a course would cause the excavation to approach the pulp too nearly. Dentistry is in many respects governed by mechanical laws, but when we come to apply mechanics to living tissues there are frequently points at which the ordinary laws must be set aside and reason allowed to hold sway. In the case of the retention of a filling, reason would set aside one law, however, only to adopt another, which, though not so general in its significance, would be indicated in a special instance. Where it becomes unwise to attempt to enlarge a cavity till its orifice is its smallest diameter, there are usually at least two directions in which extension may be made, which will sufficiently serve to hold a solid filling. Much may be gained at times by judicious roughening of the surface of the cavity. Fig. 1 exemplifies such a case.

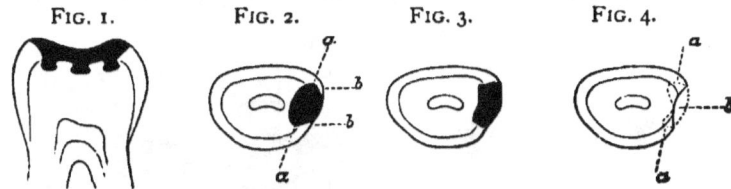

Fig. 1.   Fig. 2.   Fig. 3.   Fig. 4.

The crown of a molar has become denuded of enamel, and the dentine is highly sensitive. A few dovetails made with a sharp rose bur will serve to retain a filling, though the orifice has the largest diameter. Again, there may arise cases where the destruction leaves, not a cavity proper with an orifice, but merely a loss of substance with no retentive shape whatever. It is in these cases that the ingenuity of the dentist is taxed. In many instances well-placed screws are of great advantage. These will be described later.

In the simpler forms of cavities, those which may be described as having surrounding walls and orifices, the rule first mentioned must usually be applied, but judgment must be employed. These cavities are of three classes, approximal, crown, and surface, the latter including palatal, labial, lingual, and festoon cavities.

I shall consider approximal cavities first, because they are the most difficult, and demand more skill and judgment. Fig. 2 is a cross-section through an incisor which has been filled. A casual glance demonstates the fact that the filling could not be dislodged, because the greatest diameter, which is in the line *a, a*, is larger than the opening, *b, b*. The cavity, however, though mechanically correct, is made without due consideration of the fact that living tissue is being operated upon. If after the removal of decay the cavity naturally assumes this shape, it is permissible to fill it without further alteration; but whenever possible, it should be formed as in Fig. 3, which, while equally retentive, leaves a greater amount of dentine between the gold and the pulp. This method should be adopted in all so-called "saucer-shaped" cavities (Fig. 4), which may be deepened by grooving at *a, a,* the point *b* being left untouched.

In Fig. 5 we observe a form, which, though correct in its relation to the pulp, and formed upon mechanical principles, is nevertheless unskillfully made. It would be improper for one to make such deep undercuts intentionally, and where they have been produced by caries the orifice must be enlarged by chipping away the weak enamel, as indicated by the dotted lines *a, a*. Many fillings have failed through the well-meant but unwise efforts of the operator to give great retentive strength to his cavity by deep undercuts. Undercutting to a slight extent is imperative, but beyond that all deepening is a source of weakness. In placing gold in a cavity formed as in Fig. 5, the force used

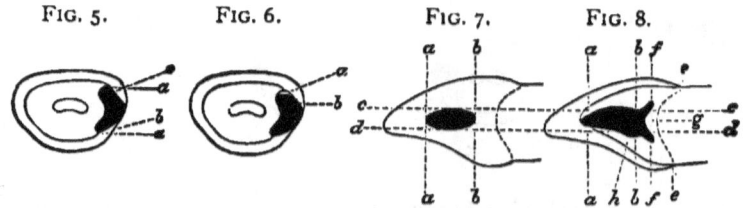

FIG. 5.   FIG. 6.   FIG. 7.   FIG. 8.

to pack the gold into the deep undercuts, whether with hand-pressure or the mallet, would tend to strain the wall outward, probably resulting in a crack, which, escaping the eye of the dentist, would yet ultimately result in the loss of a piece of enamel along the edge of the cavity. This would be more likely to occur along the palatal border, *b,* because the palatal undercut would be more accessible during the filling-operation. Thus an imperfection would be produced at a point out of sight, and the patient would be unconscious of the mishap until caries had supervened. Then we should hear our dentist say, "My filling is all right, but your tooth has decayed around it." The deep undercut at the labial aspect *c* would probably be not fully filled because of its inaccessibility, especially where the separation was but slight. Fig. 6 makes this plain. The gold does not quite reach the wall at *a*. It has been well proved that the most expert dentist cannot fill a tooth with gold so tightly that an aniline solution cannot stain the walls of a cavity. Therefore it is readily seen that as soon as the fluids of the mouth pass the slight barrier at *b* and reach such a space as that shown at *a,* we have a most favorable condition for the recurrence of caries; which explains the occurrence of the mysterious bluish color which occasionally is seen around an apparently perfect filling. If it is denied that a filling cannot be made water-tight, it is still to be remembered that in filling this labial undercut, even thus imperfectly, the strain as in the case of the palatal aspect would probably crack the thin enamel. The subsequent loss of even a tiny particle of enamel would give sufficient entrance for fluids into this artificially-produced lacuna. If, in addition to this, the excavation had not been thorough and just a little caries were left, it would require no gray-beard to explain the blueness.

In Fig. 7 we have an approximal surface occupied by a filling which thus indicates the outlines of the orifice of the cavity. Fig. 8 is placed in juxtaposition, and lines are drawn through the two for comparison. The latter is a longitudinal section through the greatest diameter of the filling, the approximal portion having been thus removed. By comparing the lines a, a in the two figures, it is seen that the undercut toward the cutting-edge is very slight. Any great extension in this direction is not only unnecessary but endangers the corner of the tooth, already weakened by decay. Comparing the lines b, b, it is seen that this portion of the cavity is extended much beyond the border. Also observe that this extension takes the form of horns, conforming in shape with the imaginary line e, e, which represents the gingival border of the enamel. By proceeding thus it is plain that we leave the enamel at the gingival border of equal strength, whereas if with a rose bur we cut a groove connecting the horns f, f, we gain nothing in retentive strength, while by thus undermining the enamel unequally we leave it weak at g. In malleting gold against this border the greatest strain will be against the weakest point, g, because of unequal resistance, probably producing a crack or fracture, which readily explains in one way the oft-noticed recurrence of caries at the gingival borders. By comparing the lines c, c, the labial borders, it is seen that a slight extension, or groove, exists along the full extent, decreasing toward the cutting-edge. The lines d, d, however, indicate that the lingual borders are identical, except at h, where there is a decided dip. Along the lingual portion of an approximal cavity (in the six anterior teeth) it is usually safer to make no groove except at the upper third, where the thickening of the tooth gives sufficient dentine for a deep dip, which will add great retentiveness to the cavity. Fig. 9 is important. It is a section through an incisor and filling, the labial half being removed. At a glance it is seen that on this plane the cavity has no retentive shape whatever. This is because on this plane, which is through the center of the pulp, extensive undercutting would either approach the pulp, or produce weak points in enamel. The gingival border at g is seen to be strong, and shaped to resist any necessary force which may be brought against it. Were the cavity shaped as seen in Fig. 10, we should have *g*, a weak point, as shown in Fig. 8, where we view the same thing from another aspect. Extension toward the cutting-edge as shown in this figure would make possible a fracture in the direction of *a*, which would result in the loss of the corner.

Until I take up special cavities in detail, it is not necessary to say more of approximal cavities, there being only a few points in connection with the incisor region which do not apply to bicuspids and molars, and *vice versa*.

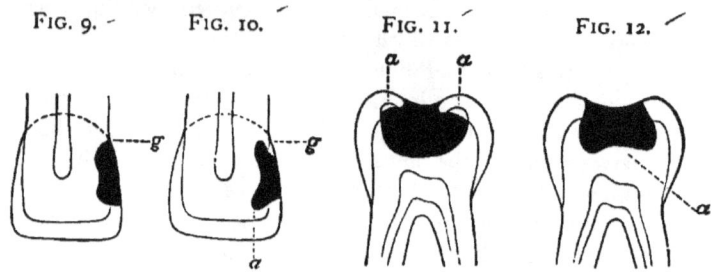

Fig. 9.   Fig. 10.   Fig. 11.   Fig. 12.

In the preparation of what are strictly crown cavities there is little to be said here beyond emphasizing the necessity already pointed out, not to leave too great undercuts. If the cavity be left as in Fig. 11 after the removal of decay, an attempt to insert gold would probably result as figured, spaces occurring at *a, a*. Subsequent mastication would crush the weak enamel, and leaky borders would ensue. Fig. 12 shows how such a cavity should be formed, the dentine being left as thick as possible (*a*) over the pulp.

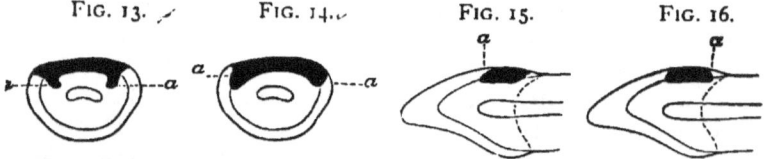

Fig. 13.   Fig. 14.   Fig. 15.   Fig. 16.

The consideration of compound cavities, involving crown and approximal surfaces, may be better discussed when I come to special cases.

Festoon cavities are those which, of all surface cavities, require most thought. Fig. 13 is a section through an incisor and filling. Grooves for retention have been made at points *a, a*, equidistant between the central line and the approximal surfaces. These being directed toward the pulp, or straight downward where the pulp is closely approached, offer counterbalancing resistance and are all-sufficient, especially because the position is one where there will be little, if any, mechanical force exerted to dislodge the filling. In this form of cavity we obtain strong edges, whereas if shaped as in Fig. 14

the edges are weak (*a, a*) and liable to fracture during the process of filling. This would especially be the case where an approximal cavity either existed or should afterward occur. In forming such an undercut it is very probable that the other cavity would be reached, and thus the enamel separating the two would be very fragile.

In Figs. 15 and 16 we have longitudinal sections, the approximal surfaces being removed. Fig. 15 shows a groove or undercut (*a*) toward the cutting-edge, but none toward the gingiva. This latter arrangement furnishes a solid surface against which to mallet gold, while the one groove, terminating in the horns pictured in Fig. 13, is all-sufficient for retention. Where a second groove is made toward the gingiva, as in Fig. 16, the enamel or border at *a* is weak, and recurrence of decay more probable.

INTENTIONAL EXTENSION OF CAVITIES.

In many cases it becomes necessary to extend the limits of cavities beyond the line of carious destruction. Where this extension is internal, and is done without enlargement of the orifice or cavity-border, the course is pursued merely to obtain proper anchorage for the filling. There are, however, times when the borders of cavities must be extended, and these cases we must consider.

It is not uncommon to find a mouth, especially among children, where the point of a fine exploring instrument will readily detect a seemingly tiny cavity in the crown of a bicuspid or a molar. Suppose the patient to be of cleanly habit, the mouth in a hygienic condition, the teeth free from stains, and the caries unaccompanied by discoloration. The operator selects a small rose bur, which slightly enlarges the opening and then sinks into the dentine, throwing out a *débris* of white decay ; to be conscientiously thorough, the case being a bicuspid, he carries the bur along the sulcus to the opposite pit. The cavity is then filled, fine-pointed pluggers and small pieces of gold being used, the whole when finished presenting a beautiful polished appearance. The crown is improved, since now it is jeweled ! Suppose that eight or ten of such narrow streaks of gold be placed in different teeth about the mouth. They ought to last a lifetime, one would think. Then why is it that a year later some leak, while others show a bluish discoloration about their borders?

The reason is that in his endeavor to save tooth-substance the operator did not make the cavities large enough. He did not remove all of the decay, and he could not do so, because, first, he did not sufficiently enlarge the opening to the carious region, and second, he prepared the cavities with a bur alone. It is frequently impossible to detect all the ramifications of decay with an engine-bur, partly because the rapid motion destroys the keen sense of touch, and more often because the position of the cavity is such that the shank of the instrument limits the territory which can be reached. All cavities should be scraped with spoon excavators, and all undercuts explored with a right-angled hatchet. For this reason all crown cavities should be opened sufficiently to permit the use of hand instruments.

Fig. 17 shows an extreme case. The patient presented, complaining that there was a dull pain in an indicated region, which she could not definitely locate in any special tooth. A second left superior bicuspid appearing discolored, an examination was made in search of a cavity. A fine explorer was passed, requiring some pressure, into an opening in the anterior pit of the sulcus. On attempting to remove the instrument the entire crown came away, much to the horror of the patient.

Close examination showed that the only defect in the enamel was the small aperture through which the explorer had been passed. Caries had its starting-point here, and reaching the dentine, the devastation continued, without further affecting the enamel, until the pulp had been reached and killed. The crown when cleansed of decay

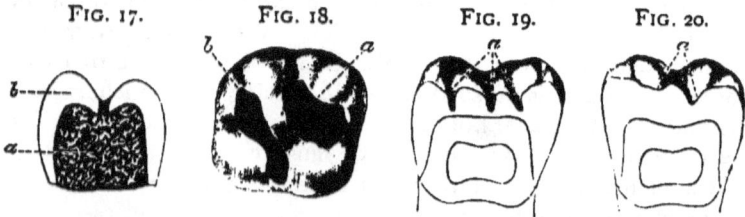

FIG. 17.   FIG. 18.   FIG. 19.   FIG. 20.

presented as a complete hollow shell or cup of enamel. In Figure 17, which shows a section through the crown, the enamel b is seen to be intact except where the probe passed, a representing the dentine, all of which was carious.

There is a lesson to be learned here. Whenever caries reaches the dentine, whether through a sulcus in a crown or through the smooth enamel on an approximal surface, further extension of decay will be in the dentine. Enlargement of the orifice is due to the fact that as the enamel loses its support from being undermined, the thinned borders chip or crumble off. This is specially true of masticating surfaces. It follows therefore that all discoloration in the sulci is to be regarded with suspicion. Dark fissures should be opened along their whole extent, sufficiently to allow proper examination with a hatchet excavator. If it is found that the enamel is undermined, the cavity can best be further enlarged with a chisel and light taps of a mallet, thus splitting off overhanging borders with the least pain, and revealing the true extent of the cavity.

Fissure-cavities, however, are to be opened even when not discolored, if a cavity is found at any part of the sulcus. It is poor practice to insert pin-head fillings in such positions. It would be safe to say that the sulcus in a bicuspid should always be opened to its full extent, even when caries appears in but one pit. There may be a few cases which might rightfully be excepted, but, as stated, the rule is a safe one. From this dogma it follows that it is improper ever to insert two fillings in the crown of a bicuspid, one in the pit at one extremity of the sulcus, and one at the other end. The patient is thus made to believe that he receives two fillings, when truly but one is needed, and would serve better than two, since the space between is weak.

In considering molars we perhaps should be less dogmatic, though in the inferior jaw we would still be erring on the safe side, if erring at all, when we cut out the crossed sulci. I heard a professor in a dental college once, in a lecture, instruct his class somewhat differently upon the point as here stated. He said that he had frequently filled lower sixth-year molars, placing one filling at each corner of the cross, and one in the center, charging five dollars apiece. Had this gentleman made a single filling of the whole, charging twenty-five dollars for his operation, he would have been more scientific, more honest, and yet have received the same fee. A man may regulate his fees as he chooses, but he cannot rightfully arrange his fillings for effect, at the expense of durability.

It is necessary here to dilate a moment upon this subject of enlarging the cavities in the lower molars, especially those of the sixth year. We are usually asked to fill them in the mouths of children. The parent has no suspicion that there is a cavity at all, and when the child returns home with a tremendous surface of gold showing, the mother calls at the next visit, charging perhaps that the dentist has "bored a hole in the tooth." This is the opportune time for educating the family. A candid explanation should be made, stating that while it would have been possible to fill the tooth with only a small amount of gold, it would surely have needed refilling in a few years. To those who must have an incentive for doing conscientious work, aside from the gratification of knowing that the most skillful course has been pursued, it may be said that these talks with parents, and patients generally, are sure to redound to the benefit of the operator.

The general rule of cutting out the sulci of course holds also with the superior molars, but here it is often best to insert two fillings. There is commonly found a deep pit in the anterior portion of the crown, a less deep one existing posteriorly, with a sulcus extending well over into the palatal surface. Fig. 18 shows the crown of a superior molar properly filled, the anterior cavity (a) being built up so as to be free from ramification of the sulci, and the posterior cavity (b) having been cut so as to include the palatal groove. This latter direction should be followed rigidly, except where the groove is rather intimated than actually existent. It will sometimes happen that the anterior and posterior pits, instead of being separated by a ridge, will be slightly connected by a groove across the ridge. In such a case two courses are open, dependent upon circumstances. Where the cavities may be made separately of strong retaining shape, they should be so formed, but in placing the gold the filling may be made continuous from one cavity to the other across the groove. This accomplishes the object with the least loss of substance. Occasionally, however, the filling would be better retained if the two cavities were united, in which case the bur or drill may be used. I prefer an inverted cone bur here, thus forming at once a slight but sufficient undercut.

I said that the prescribed rule cannot be as dogmatically adhered to in the treatment of molars as in the case of bicuspids. This is explained by the assertion that not infrequently cases present where it would be useless to extend a fissure-cavity.

Fig. 19 shows a section through a molar. Where the sulci occur the enamel is seen to have been formed separately over each cusp, there being deep pits (*a*). This class of fissures is frequently found in the mouths of children. To the eye there is no sign of decay, and there may be none, but the fine probe will be caught all along the line of the fissure, and especially where two sulci cross. These imperfections (*a*) are not seen in adult mouths, because where they exist cavities will be formed before puberty. These fissures should be opened and filled, regardless of the absence of decay, as systematically as when decay is present.

Fig. 19 is from a first superior molar, and the natural pits (*a*) are seen anteriorly and posteriorly. This imperfection more often occurs in the inferior molars.

In Fig. 20 we see a section through a superior molar which has simply rounded grooves as sulci (*a*), and the enamel is thick over the dentine at all points. This class of teeth is found in either jaw, though more commonly in the upper, and is usually associated with a strong, fully-developed frame. The teeth are large and the enamel dense. Should caries appear at one end of a sulcus, it would be unwise to extend the cavity as described. It is more than probable that decay has supervened in consequence of some slight imperfection at the special point attacked, to fill which is all that is needed.

To summarize as to crown cavities in molars and bicuspids, I would advise as a rule the placing of but one filling in the bicuspids and lower molars, cutting out the sulci fully. In the superior molars two fillings are to be placed where a marked ridge separates the anterior from the posterior pit. Where the intervening groove is fissured, the cavities are to be united. This will occur more frequently in the second than in the first molar, and more often still in the third molar.

I have advised extending the crown cavity into the palatal groove of the superior molar, but in these cases the posterior fissure is usually markedly continuous with the palatal groove. With the buccal groove it is different. In the inferior jaw this groove often presents with its widest part toward the gum, narrowing as it approaches the crown. If this groove is non-carious, it would usually be unwise to extend a crown cavity into it; if slightly carious, it is generally good practice to prepare the cavity pear-shaped, the apex toward the crown, and not to actually connect the crown cavity with it. Where caries is extensive at both points, they must be united, but even then it is best to unite them as little as possible, as by deep undercutting, the buccal walls become very much weakened. Prepare both cavities as though they were to be filled separately, and then pass a bur along from one to the other, making as shallow a channel as is compatible with strength.

Where a crown cavity in a bicuspid or a molar is but slightly separated from an approximal cavity, the two should be united and filled as one. If filled separately, the frail enamel partition must eventually be crushed out under the force of mastication.

There is a kind of extension which is unfortunately too often necessary. A patient presents to have the teeth cleansed. In removing stains along the gum-line of a molar, the enamel is found to have become softened. Examination with a hatchet excavator demonstrates that there are two, or perhaps three, small cavities in close juxtaposition. The rule here is rigid. The cavities must be opened up and formed into a single groove; moreover, it must be extended as far as the engine-bur will cut easily. All softened or softening material must be cut away until firm margins are reached. I have known such procedure to involve the removal of nearly all of the buccal surface, and once I so formed a cavity completely encircling the tooth near the neck, but, as I have said, this is a rule which must invariably be enforced. A somewhat similar condition, though from a different cause and different in character, may arise from the use of a clasp-plate. Here extension is indicated in another way. In the first instance, we aim simply to reach a portion of tooth-substance which is strong. In the second, this would not suffice. It will frequently be found that the caries exactly marks the part of the tooth underlying the clasp, the cavity cleaned of decay showing just the form of the clasp. Such a cavity should be extended in all directions, so that when the plate is again placed in position the clasp will not touch the tooth, but will rest against the filling. If this course is not pursued, it is plain that the width of the clasp and of the filling being the same, the edge of one lies just along the border of the other. As all accumulation of food along the edge of the clasp would of necessity be just along the margin of the filling, it is easy to understand that leakage would soon ensue.

We have now to consider the anterior teeth. In this region a new condition enters which must often modify our methods. What would be good practice in an inconspicuous part of the mouth might be very reprehensible in the front of the mouth, because of the resulting disfigurement. This exception to general rules must be further modified by sex, for we might not hesitate to enlarge a cavity in the mouth of a man wearing a heavy moustache, where we should regret exceedingly to place a large gold filling in the white teeth of a handsome young miss.

The central incisors occasionally are found grooved and pitted near the cutting-edge. Suppose that two or three of these pits are found to be actually carious. Shall we unite them by cutting out the groove, as in the case of a molar? Most assuredly not, for by so doing we greatly disfigure the mouth, besides materially weakening the end of the tooth. This class of decay is seen in Fig. 21, and the three cavi-

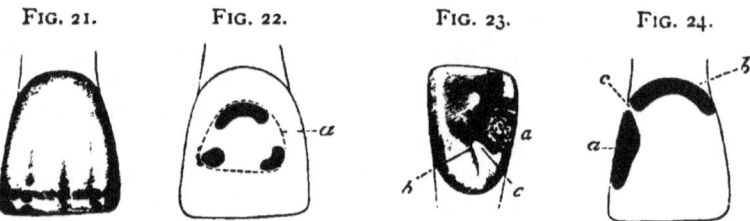

Fig. 21.   Fig. 22.   Fig. 23.   Fig. 24.

ties shown should be filled separately. The central incisors are usually longer than the laterals, and if this extra length is sufficient to allow grinding off the imperfect end, thus removing entirely the pits and groove, it should be done, provided the centrals are not thereby made shorter than the adjacent teeth.

Fig. 22 shows another condition. The tooth, originally perfect, has become decayed as a result of the presence of green-stain. When the stain is removed, we find, let us suppose, three distinct cavities. The dotted line a in the figure indicates the smallest cavity uniting the three which could be made. The result would be a deplorable disfigurement. To unite the two smaller cavities, thus placing two fillings, would be almost as bad. It will therefore be preferable to fill the three cavities separately. Of several evils we must choose the least.

The six anterior teeth are sometimes seen with distinct grooves in their palatal surfaces. This occurs most frequently in the lateral incisors, and least often in the cuspids. Of course where such a groove is found with caries at the deepest point, it is essential to cut it out thoroughly. Sometimes there may be no caries in the groove itself, but an approximal cavity closely approaches it. This is shown in Fig. 23, which gives the palatal aspect of a lateral incisor. The palatal border of the approximal cavity, *a*, closely approaches the end of the sulcus, *b*. If we fill the cavity without taking note of the sulcus, it is possible that no future harm may supervene. It is more probable, however, that caries would occur in the groove, and burrowing under the filling would loosen it. It might be claimed that when this occurs would be the time to attend to the case, but it must be remembered that when a tooth has been filled, the generally accepted idea of the patient is that it is safe as long as the filling remains in position. The final loss of a filling by the burrowing of decay beneath it may be delayed for an indefinite period. Thus caries would be progressing without being suspected until a sudden pain gave warning, and by that time the mischief would have been done, the pulp being either exposed or closely approached. It is therefore preferable to fill the groove when treating the approximal cavity. The method of procedure is slightly different from that which would be pursued for a molar. If we unite the cavities at the outset by a fairly deep groove, we would, especially in a lateral incisor, produce a very weak point at *c*, which would be likely to crumble under the mallet during filling. The better plan, therefore, is to fill the approximal cavity first. This being done, prepare a cavity along the sulcus decreasingly deep as you approach the approximal filling, just reaching that point and no more, making it as shallow as would be consistent with strength.

The treatment of a somewhat similar condition upon the labial surface must be quite different. Fig. 24 shows an incisor having an approximal cavity (*a*) and a festoon cavity (*b*). The space (*c*) between is necessarily a weak point, and would be further weakened in preparing the two cavities for the reception of separate fillings. It is much better to fill them as one cavity, cutting away entirely the intervening enamel and dentine.

We now come to a most important subject, approximal cavities. Are we under any circumstances to extend such cavities beyond the borders outlined by the decay? There are certain cases where it is imperative, and others where such a course should be prohibited.

In the six anterior teeth the approximal cavities should be prepared as small as possible, consistent with reaching strong edges. I am aware that this is a dogmatic assertion and at variance with teaching of high authority. Nevertheless my experience teaches me that this course cannot be too emphatically insisted upon. Extension of approximal cavities in incisors has been advised, in two directions. First toward the gingiva, to reach a point anticipatory of the possible recession of the gum; and second around the palatal and labial angles, to reach self-cleansing lines. As to the first proposition, the recurrence of decay toward the gingival borders of a filling, in my opinion, depends not upon the position which the border occupies, but rather upon the form given to the border, the manner of insertion of the filling, and very largely upon the perfection of the finish. As discussion of this must occur at a later point in this work, no more need be said here. As to the second claim, that of reaching self-cleansing surfaces, this will be as good a place as any for determining the point. What is a "self-cleansing" surface? Plainly the term is a misnomer. *No surface can cleanse itself.* Manifestly it is meant to imply a surface readily cleansed by the tongue or lips. The only surface of the former character is the lingual aspect of the six lower anterior teeth. That the tip of the tongue does serve the purpose here, is amply proven by the fact that caries seldom if ever is seen in the lingual surface of a lower incisor, *the only point on any tooth where I have never placed a filling.* I have placed distinctly lingual fillings (not complicated with approximal cavities) in lower cuspids, but never in an incisor. This would limit this saving, cleansing power of the tongue to a single surface, of only four teeth. Then when we remember that this very point is the first place we attack in cleansing a set of teeth, we see what limited ability as a scavenger must be attributed to the tongue. Next we have the lips. It is rare that we observe a cavity in the labial surfaces of incisors or cuspids except along the festoons, under green-stain, or as a result of erosion, in which latter case it is most probable that the lips, or the mucous follicles on their surfaces, produce rather than retard the destructive process. With these exceptions, we must admit that the lips may effect a saving cleanliness. We find that festoon cavities are quite common. This indicates that the farther we get from that part of the tooth—viz, the cutting-edge—which can be reached by the edges of the lips, the less cleansing we observe. *If the lips cannot cleanse along the festoons, it is evident that they cannot, and do not, cleanse around the curve toward the approximal surfaces.* Can this be done by the tongue? It is rarely if ever that one washes the labial surfaces of the lower teeth with the tip of the tongue; and as to the upper, though this is more common, the action is to place the tip of the tongue about the bicuspid region, carry it along toward the bicuspids of the opposite side, and then back. This would wash off the *débris* along the labial surfaces, *but would crowd it between the approximal curved angles*, so that if we are to extend approximal cavities around this angle to reach an imaginary cleansing surface, the extension should be very great. Fig. 25 shows a section through two incisors having approximal fillings as I should advise them to be placed, the borders not having been extended either lingually or labially. Fig. 26 shows a similar section, the cavities having been extended to what are described as "self-cleansing lines," represented in the figure by the lines a, a, b, b. These lines indicate the surfaces touched by the tongue as it is passed over the teeth. Observe that the extension of the cavity-borders in the figure has been carried beyond the points of actual contact between the tongue and tooth (c), for it is plain that to make a border exactly upon this line would be rather to invite decay than to prevent it. A comparison of these two figures shows to what a considerable extent extension of

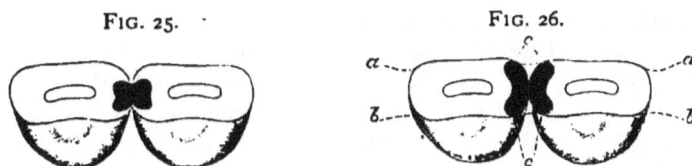

FIG. 25.   FIG. 26.

the cavity-borders must be carried to reach these imaginary "self-cleansing" surfaces, and I argue that when we consider the limitations of this tongue- and lip-cleansing, as a tooth-saving influence, compared to the manifest disfigurement of the mouth by the greater display of filling-material, the comparative infrequency of recurrence of decay along these borders where gold has been properly placed, and more than all the lessened retentive power of the cavity, or greater approach toward the pulp, it follows that to extend an approximal cavity in the six anterior teeth in any direction beyond

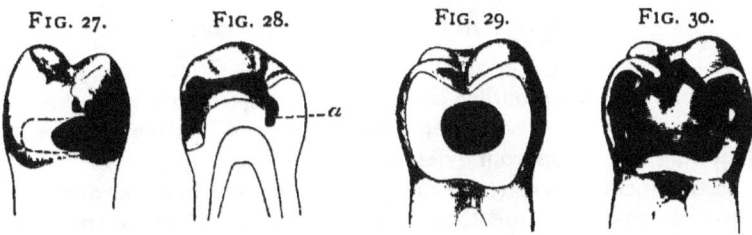

FIG. 27.   FIG. 28.   FIG. 29.   FIG. 30.

the amount required for reaching a strong enamel-border is extremely hazardous and unwise.

Extension of approximal cavities in bicuspids and molars is a different subject. Where such a cavity occurs in a bicuspid or molar distant from the masticating surface, it being either impossible or unadvisable to obtain sufficient separation, it is better to extend the cavity so as to gain ready access than to attempt to fill it with little space. The argument in these cases that such extension must be toward the palatal rather than the buccal surface, is more often written than is the method practiced. It is preferable to extend toward the buccal surface, so that the operator may see his work. Fig. 27 shows such a cavity in a bicuspid, and the dotted line indicates the extension advised.

Where the approximal surface is decayed, presenting a large saucer-shaped cavity of a form most difficult to make retentive without encroaching upon the pulp or producing weak surrounding walls, it is wise to extend the cavity toward the crown and well across that surface, following the sulci and making a deep pit at the opposite end. Fig. 28 shows a section through a tooth so prepared, *a* being the pit depended upon to retain the filling.

A very serious condition is occasionally seen, which has been produced artificially. Under the delusion of saving a tooth from the necessity of being filled, some dentist has filed a "V-shaped self-cleansing (?) space" between two teeth, and then inserted a flat, or flush, filling such as is shown in Fig. 29. The patient complains of constant pain after meals, caused by the impaction of food into this "self-cleansing" space, and the removal of the meat-shreds with a tooth-pick, the constant use of which has brought on a highly congested state of the gum. It is plain that the only remedy is to insert a contour filling. This cannot be done in the small cavity surrounded by thick flat-faced walls, as figured. Therefore the cavity must be extended till the walls are thinned at their edges, assuming the shape shown in Fig. 30. A properly contoured filling may then be made with the assurance that there will be no further complaint after the gum has healed, which will occur very soon.

Formation of Cavity Borders.

In considering the edge-line of cavities, I shall make a distinction between the general outline and the actual edge, nominating the one "cavity border" and the other "enamel margin." Both are subjects requiring discussion from different stand-points.

As a general principle, the cavity border must be free from angles,—must be curved. More than this, it should be a continuous curve rather than a succession of curves. From an esthetic stand-point, nothing is more essential than that the eye should rest upon a gently curving line, rather than upon an angular or scalloped one. Many operators in their efforts to economize tooth-substance make the mistake of producing results which are unsightly. A small filling with an odd border is more conspicuous than one which may be larger, but which has been placed in a cavity correctly formed. These rules, of course, apply more especially to such fillings as reach the labial surfaces of the anterior teeth. They apply, however, to the molar region also, because in this instance esthetics and durability go hand in hand. To leave irregular borders in any cavity is to produce weakness along the points of contact between the tooth and the filling. Chipping of the margin will occur either during the placing of the filling or later.

Reference to a few special cases will demonstrate the point more fully and clearly. Fig. 31 is diagrammatic, but conveys the idea clearly enough. It is supposed to represent the ragged outline of a cavity in the approximal surface of a central incisor. In Fig. 32 we observe the same, prepared and filled, according to methods too often followed. There are three errors along the border. The cavity has been formed on curved lines, it is true, but they are weak lines and not esthetic. The slight prominence at a has been left, thus producing an undulating border, instead of the more beautiful curve seen in Fig. 33. Moreover, if we study the cleavage of enamel we discover that this prominence is weak, since it is unsupported, and must almost certainly crack during the operation of filling. The force exerted in packing gold against it would probably produce fracture along

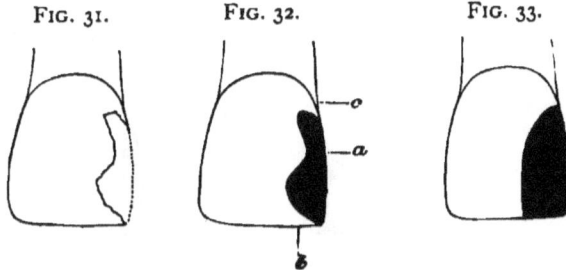

FIG. 31.   FIG. 32.   FIG. 33.

the line of the enamel-rods, and if the disengaged piece did not come away at once it would do so later, or else decay would be invited here. For similar reasons the sharp point at b should not have been left. Even if no fracture were produced during the packing of gold, the loss of a triangular segment would invariably follow upon use in mastication. Again, the heavy undercut at c is an error, for the strongest gingival border is one formed without undercut of any kind, but having a flat surface against which force may be freely exerted without fear of chipping. Fig. 33 is correctly formed and decidedly more pleasing to the eye. In Fig. 34 we have another cavity, which is too often incorrectly formed, as seen in Fig. 35. This should have a border similar to Fig. 33. As prepared in Fig. 35 the point a is weak. Gold is a soft, malleable metal, and where it is brought to a thin edge, as seen, the impacting force of mastication tends to flatten it out until it curls and breaks, leaving a ragged edge. The point b is weak for reasons already given.

As I have made my diagrams to deal with cavities involving the restoration of corners, this will be an opportune place to discuss that particular subject. It has been recommended by recognized authority to form a cavity of this character as shown by the heavy shading in Fig. 36. To my mind this is most unfortunate teaching. We thus get two sharp angles, both of which must become weak points save in the hands of the most skillful, even if this exception be allowable, which is doubtful. In any event the result is far from beautiful. Such a cavity coming into my hands would have been formed as shown in Fig. 33, for the L-like extension is supposed to have been intentionally made by the operator. Once I was called upon to refill a tooth which had been operated upon according to this method, and in that instance I formed my cavity as indicated by the dotted line, Fig. 36, approaching my ideal as nearly as possible under the circumstances. I could have shaped it as seen in Fig. 37, but such a form is so uncommon as to be more conspicuous than that chosen. In Fig. 38 we see a diagram showing two corners contoured. In cases of fracture

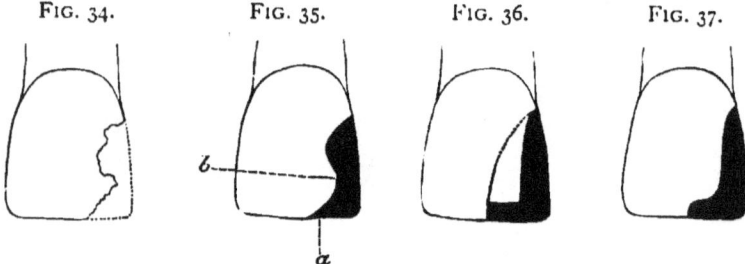

FIG. 34.   FIG. 35.   FIG. 36.   FIG. 37.

from falling against a curbstone or from other accident, the corner will usually be lost, the line of fracture being a straight one. A filling placed without alteration of the border would appear as shown at a, and the correct border is seen at b. Let us consider these two forms aside from esthetics. Suppose that the dotted lines f, f, give the internal retentive shape of the cavities. The force during mastication, which will tend to dislodge the fillings, will come in the direction indicated by the dotted lines c, c. The border of the corner a being a straight line, it is evident that the retention of the filling, under strain, will be entirely dependent upon the strength of the wall at d, of course plus the resistance of the lower undercut. In the other corner (b) we have equal strength, plus the resistance offered by a well-defined shoulder at e. So that once more we find the curved line stronger, as well as more attractive in appearance.

In cavities such as are shown in Fig. 39, and even where the depredation has been less, it has been advised to remove the natural corner as far as the dotted line *a*, and place a contour filling. The argument is, that the tendency toward cleavage in both enamel and dentine is so great that this corner must be lost sooner or later, so that it is best to fill at once as directed.

This again I think a grave error. Many deductions have been made as to the cleavage of enamel and dentine which are misleading, because the experiments have been conducted upon dried teeth, out of the mouth. Using the chisel upon these, the investigator finds that very little force need be exerted, either to separate parts of enamel along the line of the enamel-rods, or to cleave off slabs of enamel from the underlying dentine. In consequence of this latter fact one gentleman has advised that where the border of decay approaches nearly the gingiva, the narrow remainder of enamel should be chipped away and the gingival edge of the gold allowed to rest against the cementum. It seems to me that this is going to a great and hazardous extreme in following a theory.

The truth is that we do not have to deal with dried teeth in our practice. Therefore if we consider cleavage at all we must study it as

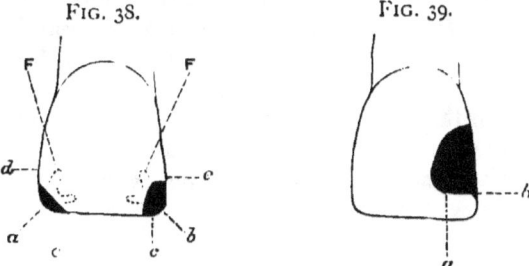

FIG. 38. FIG. 39.

we have to contend with it. We are called upon to fill two classes of teeth,—those having living pulps, and those which are partly devitalized because of the loss of that organ. A wholly dead tooth, one in which both pulp and pericementum have died, should invariably be extracted.

In deciding as to the form to be given to such a cavity as we are discussing, it becomes important to determine whether the pulp is alive or dead, and if dead, whether it has been freshly devitalized, or, having died from disease, the tooth-substance has been permeated by putrescent matter.

Supposing that the pulp be alive and the tooth in a state of health, I consider that it would be malpractice to remove the natural corner, and my advice is based upon the result of practice rather than upon theories, which, though apparently logical, often prove unpractical.

The cavity borders should be curved as gracefully as possible without making further encroachment upon the tooth-substance, and the interior so arranged that the filling will be retained from above, and laterally, no dependence being placed upon the weak corner. A slight bevel at *b*, Fig. 39, will possibly serve as a point of resistance against such force as would have a tendency to crack the corner; at least it is preferable to a sharp angle. At the time of so filling a tooth, it is as well to explain to the patient that the corner is weak, and may be broken off in the future; that you are obliged to choose between taking it off and risking future accident; that you follow the latter course to give her the benefit of the doubt and save her from the disfigurement of a larger filling as long as possible; and finally that you have so arranged your filling that if the mishap should occur you can build on more gold, *without removing what you have just inserted.*

By this course it is seen that nothing is lost, whereas, should the fracture occur, *the patient will even then be in no worse predicament than that in which her dentist would place her by the intentional removal of the corner;* if the fracture should not supervene, she would certainly be the gainer. It may be observed that I have alluded to the patient as "she." This is because the force of my argument is greater when dealing with women, who do not wear moustaches, and the marring of whose beauty means more than does that of a man.

Thus it follows that in healthy teeth, whilst I recognize the element of the cleavage of enamel, I have also noted the remarkable strength which may exist in an apparently weak spot, due to the support of healthy underlying dentine, and the tenacity with which the two cohere.

Where a pulp has been freshly devitalized, the tooth being otherwise healthy, the fact that the root-canal offers such excellent opportunity for anchorage that it would be absolutely certain that future damage could be repaired without removal of the initial filling, would lead me to follow the same method.

Where a tooth comes into my hands in which a pulp has been long dead, the substance of the tooth discolored and evidently friable, it then becomes a subject for special consideration. In the diagram I have shown the limit at which I should hesitate. Where the decay had proceeded farther so as to make the corner weaker than here shown, I, probably, should remove the corner. To determine whether to do so or not, place a *dull* instrument against the corner and exert pressure. If cleavage occurs under such strain, the dentist need have no doubt that he has followed a proper course. A few gentle taps with a mallet might even be resorted to, and if a crack should appear, the corner should be removed.

There is one other condition where it is not only advisable but im-

FORMATION OF CAVITY BORDERS.

perative to remove the corner. Having determined not to do so, suppose that the operator proceeds to pack his gold. Reaching the weak spot, he abandons the mallet and uses hand-pressure. In spite of this precaution, a crack suddenly appears along the line of cleavage. The corner should be removed at once, whether the tooth be healthy or otherwise.

To emphasize the above arguments, I may state that an examination of my records shows that I have filled hundreds of such teeth in the manner here advised, and that in only two instances have the corners broken away afterward. In both cases the contour was restored without removal of the filling. I have also a few times repaired broken corners without removing the original filling, where the patients had been in other hands before coming to me.

The economist of tooth-substance frequently makes a mistake in forming cavities in the approximal surfaces of molars or bicuspids.

FIG. 40.   FIG. 41.   FIG. 42.   FIG. 43.   FIG. 44.

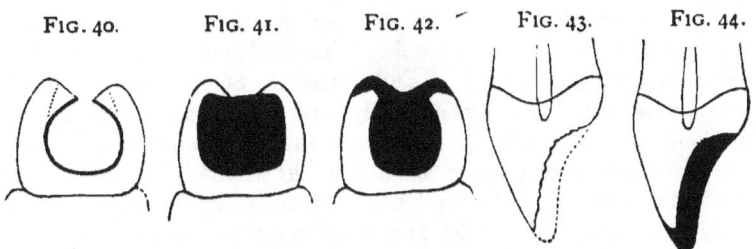

In such a cavity as is shown in Fig. 40, he shapes the cavity as indicated by the heavy line. Operating between separated teeth, with the little space which we usually have, it is plain that he is at a great disadvantage as soon as his filling is half completed. He is then obliged to fill places which it is almost impossible for him to see. It would be much better to extend the border to one or both of the dotted lines, as might be required to gain ready access to the cavity at all parts.

In the masticating surfaces of molars no special directions are necessary, beyond advising the use of a rose-bur along the sulci, large enough to allow the entrance of a plugger if gold is to be inserted, and larger still where amalgam is to be used. It is a mistake to cut out sulci with small burs, and then fill with amalgam. The opening all along the line should admit a small ball-burnisher, so that in packing the amalgam the mercury may be thoroughly expressed. Many amalgam fillings have failed because of tiny burs having been used at the extremities of the sulci, so that when placing the amalgam these points were improperly filled, either with material having an excess of mercury, which on that account has never set, or with an insufficient quantity dragged or wiped into the crevice with the cotton, spunk, or bibulous paper used for smoothing the filling. If a fair-sized rose-bur is used, all corners will be well rounded and accessible. It will sometimes happen that by following two nearly parallel sulci to their extremities we leave between a narrow strip, vhich, if left, is apt to crumble under the blows of the mallet. In such cases the intervening bit should be cut out, and the two sulci thus united.

In a few rare instances the border of a cavity may be so placed that a filling becomes a support to a frail wall rather than dependent upon it. A case from practice will well explain this. A patient once came into my hands having a first bicuspid which was positively black. This was before the days of the perfection of artificial crowning. Examination showed the tooth to have been filled with amalgam. The cavity occupied a large space directly through the tooth from the anterior to the posterior approximal surface. Fig. 41 shows how the dentist had formed the cavity. It will be seen at a glance that the filling offered no support whatever to the weakened walls. On the contrary, in mastication, a piece of food acting as a wedge between the cusps would have a tendency to drive the labial and palatal walls apart and away from the filling. A slight elasticity of the dentine allowed this. Thus the filling leaked badly, and discoloration was largely due to decay between the amalgam and surrounding dentine. Great care was required to remove this filling, but when accomplished, and all the decay removed, the tooth was far from being unsightly in color. I filled this tooth with gold, anchoring the filling in the root-canal, and so shaped the borders that the filling became a strong and complete support to the walls. This is shown in Fig. 42. It is plain now that all mastication must be upon the gold. The ends of the cusps have been removed and the upper extremities of the remaining walls strongly beveled, so that the gold built over them holds them firmly from any tendency to bulge outward.

Similar treatment is required in cases shown in Fig. 43, where abrasion against the lower teeth has worn away the palatal surface of the upper incisors or cuspids, leaving frequently a knife-edge. This incising-edge should be ground down to a well-marked bevel and gold contoured over it as shown in Fig. 44.

ENAMEL MARGINS.

A great deal has been written and preached upon this subject, and, as in other matters, the theorist has advanced several erroneous propositions. The desideratum in all fillings is that when polished the line of contact shall be as fine as a hair. Examination with a jeweler's lens should not reveal any marked raggedness. Considering the arrangement of the margin from this aspect alone, there is but one method of producing the result desired. The margin must be smooth, and at a right angle to the outer surface of the tooth, in all cavities save those in masticating surfaces. These latter will be discussed separately.

In Fig. 45 we see a section through a central incisor and contoured gold corner. The margin at the labial surface *a* is a sharp right angle, as is also the margin on the palatal side. In filling we of course lap our gold well over the margin, and then trim away, seeking a sharp edge line. Any method of polishing which scrapes *across* the margin will deceive as to the true edge. That is to say, it is an error to use stones, polishing wheels, or sand-paper in disks or strips, revolving the same around the tooth from one approximal surface to the other. This method may be employed at the outset to remove the excess of gold; but when it becomes necessary to polish away the last remaining overlap, in order to reveal the true margin and leave a beautiful fine line, the polishing must be done along the length of the margin, from the incisive edge toward the gum. If it is an approximal cavity I prefer fine sand-paper disks, pliable enough to bend read-

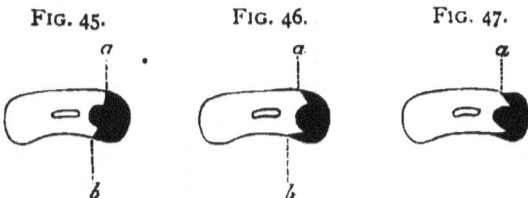

FIG. 45.   FIG. 46.   FIG. 47.

ily, though stiff enough to cut well. Revolving the disk rapidly as described, the edge of the disk is made to approach the edge of the gold; if the latter is overlapping the margin, the disk attacks it instantly and removes sufficient to perfect the edge, leaving gold and tooth-surface exactly flush. The same result may occasionally be expected from the use of a fine polishing-stone (Arkansas or Colorado stone, not corundum), but it must be fine enough not to cut enamel, and must cut gold very slowly. Margins of fillings along the palatal angle may best be polished with fine disks, but where the border reaches the concaved surface small stones revolving from incisive edge toward the gum best attain the finest finish. A stone turning in this direction will not catch in the dam.

If approximal margins be beveled outwardly (Fig. 46), as some contend is requisite, it is plain that we no longer have a really defined margin at *a* or at *b*, but simply an obtuse angle along a receding slope. If this were really an obtuse angle, it would be bad enough, but it is almost certain that the operator will produce a rounding bevel, so that there is really no distinct margin at all. Now if gold be lapped over this, how shall it be polished so as to obtain a strong edge to the filling and a neat, sharp marginal line? The answer must be, that it is impossible. The cross-motion of a polishing instrument will not always expose the true edge even where the margin is a sharp right angle, and it is less apt to do so along a bevel. If the disk or stone is used as has been described to be the best method (revolving along the margin), the first effect will be to produce a scalloped line, and if the operator then attempts to straighten out that line, he removes the edge of the gold, so that a portion of the beveled enamel is uncovered, leaving a slight groove along the length of the tooth next to the filling. Fig. 47 shows a section through a tooth and filling after such a mishap, the exposed bevel resulting in a groove showing at *a*.

This idea of beveling is much more frequently advocated at the gingival border, through a strange impression that the margin which has gold lapping over it will better resist caries than where the gold is brought against it in sharp contact, without overlapping. Where such a beveled gingival margin is polished with either sand-paper strips or the approximal trimmer, the groove described is most likely to occur. If the disk alone is depended on, the overlapping gold will be left with a ragged edge in ninety per cent. of cases. If the margin along the gingival border has been formed with a sharply-defined angle, and the enamel is supported internally by dentine, which has not been erroneously cut away for anchorage, then gold may be packed solidly against and over it, and in polishing the approximal trimmer may be freely used until it no longer catches, when a fine sand-paper strip will finish the marginal line as fine as a hair.

It has been wisely argued that the enamel margin should not assume an acute angle, for the reason that the enamel at the extreme edge being unsupported by dentine, will be likely to crumble during the operation of filling, because of its tendency to cleave. For this reason it has been advised that the enamel margin should assume the direction of the enamel-rods at the point operated upon.

Thus theory as well as practice, and mechanics, recommends that the enamel margin should cut the surface of the tooth at a right angle, for it is usually true that the enamel-rods are arranged perpendicularly around the dentine, so that they must assume almost if not quite right angles to the surface.

If we examine the varying surfaces in the masticating portion of a molar or bicuspid, it becomes plain that to follow the rule here and arrange the enamel margin so that at all points it shall be a right angle, would be to undertake the impossible. Neither is it necessary, for whereas in approximal and surface fillings we are compelled to finish the gold down flush with the true margin of the cavity, in molars and bicuspids the sulci being depressions which we wish to eradicate, as well as to simply fill the cavity proper, we customarily extend the gold beyond the true cavity-edge, partly obliterating the natural depressions. If a filling so placed is well burnished and any ragged edges trimmed off, the form of the margin need give little concern, since it is entirely covered. Overlapping in masticating surfaces is not objectionable, because, whilst the gold may spread out, here as elsewhere, it is in such a position that all the force exerted simply presses it more tightly against the tooth-substance which it is intended to cover and protect.

CHAPTER II.

GENERAL PRINCIPLES INVOLVED IN THE FILLING OF TEETH—METHODS OF KEEPING CAVITIES DRY—THE RUBBER-DAM—LIGATURES—CLAMPS—LEAKAGE—THE NAPKIN—CHLORO-PERCHA—WEDGES *vs.* SEPARATORS—THE USES AND DANGERS OF MATRICES.

METHODS OF KEEPING CAVITIES DRY.

DURING the insertion of any kind of filling-material, all cavities should be kept dry unless this is absolutely impossible, which rarely occurs. It does not follow from this axiom that the rubber-dam is invariably necessary. There are very many occasions when a dexterous operator can dispense with the dam, which is always annoying to the patient, and yet insert as good a filling. As a conspicuous illustration of this, I need but to cite cavities occurring in the anterior sulci of superior sixth-year molars. Where the patient is a child, and the cavity small, it is unwarrantable to force a clamp over a short tooth-crown, crowding painfully against sensitive gum-tissue, with the risk of a slip during the operation. This would necessitate the removal of the half-finished filling, and beginning it anew, with the gum bleeding, thus adding to the difficulties of an operation which could have been performed with little trouble by using a napkin. Another condition where the dam is sometimes contraindicated is where the cavity in an approximal surface extends below the gum line. In order to expose the gingival margin, various methods have been advised, to which I need not allude at this time. Where the depredation has been quite extensive, we not infrequently meet a condition where the gum about the edge of the cavity is soft, yielding readily under pressure, but is firm and unyielding around the rest of the circumference at the neck of the tooth. This is especially true of molars. In these cases, though we may succeed in forcing the gum away from the gingival border of the cavity, we have a condition similar to that shown in Fig. 48. Here it is noticeable that while the cavity border is exposed throughout, the pedicles of gum have not yielded at the buccal and labial angles, but retain a normal position. The gum being firmly attached at other points, it is evident that if the rubber-dam be stretched over such a tooth, it will in consequence of the pedicles *a, a* span across, and occupy the line *b, b,* so that the cavity border will be below the dam. In many cases it is not only impossible to force the dam down with a ligature and keep it there, but the effort to do so, by causing hemorrhage, compels the abandonment of the operation at that sitting. Even where one succeeds in pressing the rubber and ligature below the border, it rises again when tied, because becoming taut it must assume the straight line *b, b;* if it is not made taut, leakage will occur. To overcome this difficulty, it has been suggested to use copper wire as a ligature. In a few mild cases this works very well. The wire ligature is applied about the neck of the tooth, and the ends twisted till tight. As with the silk or flax, it follows the line *b, b,* but with a suitable instrument it may now be forced below the border of the cavity, carrying the dam with it, and, being metallic, it will retain its position.

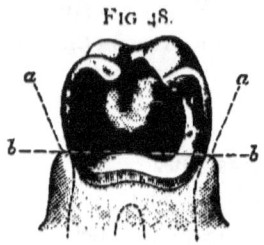

Fig. 48.

Where this cannot be accomplished, it becomes impossible to fill the entire cavity with gold. Two courses are open,—either amalgam alone must be depended upon, or amalgam may be used up to, and slightly above the line *b, b,* the rest of the cavity being filled with gold at a subsequent sitting. In either case a margin more nearly perfect will be obtained if the napkin is used instead of the rubber-dam, the gum being occasionally mopped with bibulous paper or spunk during the operation. This will be more specially alluded to in discussing the manipulation of amalgams.

When it has been decided to employ the dam, it is requisite that it should be so placed that the patient may suffer least, and the operator have the greatest facility for obtaining access to all parts of the cavity. The first desideratum is that a sufficient number of teeth should be embraced so that all folds of the rubber shall be well out of the way. There should be at least two teeth on each side of the space between the teeth which are to be filled. This as a general rule will answer in most cases. It sometimes happens, however, that much will be gained by embracing five, six, seven, or even more teeth. Cavities in the posterior surfaces of bicuspids and molars are usually trying cases, wherein accessibility and good light are essentials best obtained by stretching the rubber over several teeth. It is usually sufficient to include the central on the side of the mouth where the operation is to be performed.

METHODS OF KEEPING CAVITIES DRY.

All attempts thus far to produce a rubber of light color result in destroying its elasticity to such an extent that the material becomes unfit for dental purposes. Thus we are compelled to use a dark-colored dam, which has a decided disadvantage in that it does not reflect light. It therefore becomes important to economize the light as much as possible, by not having any unnecessary creases or folds. As the jaws are curved, it is plain that holes punched in the dam in a straight row, and by guess as to distance apart, will not permit the dam to be stretched over the teeth and lie smoothly. To obtain the proper positions for the holes lay the dam over the teeth, stretching it so that the incisive ends show through; then make a mark with an excavator, which by slightly scratching produces a whitish dash over each tooth. Punch the holes with a sharp instrument that will cut clean round holes. By this method the dam may be made to adapt itself without a wrinkle. The holes should be graded in size so that they will tightly grasp the tooth-neck, but they should not be made so small that a tear may occur at the slightest touch, or while passing over a tooth. Great care should be used to have the distances between the holes exactly right. If too narrow, they stretch to span the gap, leaving a bit of gum uncovered near the neck of one of the teeth. If too wide, especially at the space where the operation is to occur, trouble may be found in forcing the excess between teeth in close contact, or annoyance will occur from the fold produced by it. Where teeth are close together, a piece of waxed floss should be passed between all of them before any attempt is made to apply the dam. If the floss will not pass, neither will the dam. There may be a rough filling, a fracture of the enamel, a ragged edge of a cavity, or a bit of tartar, either of which will tear the dam. A safe-sided saw should be passed between these teeth, and the obstruction removed. Occasionally there is adequate space near the neck, but, for example, two incisors are in such close contact at their corners that neither ligature nor rubber can be passed between them. In such cases, drive a wedge of wood at the neck till the teeth slightly separate, when the dam may be placed with facility; then remove the wedge. It is frequently serviceable to soap the dam about the holes.

It very rarely happens that the dam can be forced between the teeth with a ligature, save perhaps in the bicuspid region, where the slanting sides of the cusps favor the method. The too frequent result is, that the ligature cuts the dam, so that while it passes into the space the operator has to contend against a disagreeable leak afterward. Where such an accident does occur, it is far better at the outset to remove the dam and apply a new piece.

When the cavity to be filled is at the labial festoon, especially where there has been a recession of gum-tissue, and caries has encroached upon the root, it is necessary to make wider spaces than ordinary between the hole in the dam which is to encircle the particular tooth and those on either side of it. If this precaution is not taken, when the clamp is used, and the dam forced up above the cavity edge, the rubber on each side will have stretched so that the gum between the teeth will not be covered, and an annoying leak will ensue. A little practice easily teaches one how to regulate the spacing in these cases. In connection with festoon cavities, the gum may be forced back and away from the edge of the cavity in either of two ways. If it can be placed so that it will have a tendency to slide under the gum, a ring cut from rubber tubing may be placed to encircle the neck, and allowed to remain in position for twenty-four hours. In this time it will be found to have forced the gum slightly back, and also by forming a space between the gum and tooth the free margin is left less resistant to pressure made by the clamp. Where the rubber ring would have a tendency to slip down toward the incisive edge, it cannot be used. In these cases a roll of cotton should be wrapped around the neck and forced under the gum as far as possible. A ligature of flax thread tied just below it will hold it in place, and in one day the swelling of the cotton will have accomplished the purpose. Either of these methods is less cruel than an attempt to force the gum up by use of a clamp.

Where a mouth is small, or the lips non-yielding, considerable difficulty may be found in placing the dam over posterior teeth. If the operator has an assistant, his aid is invaluable in placing the clamp, but in lieu of such help the patient may be made to do service. Suppose that the tooth be in the lower jaw; the operator with the second finger of his left hand can readily keep the dam from sliding up along the buccal side of the tooth. Momentarily he has the second finger of his right hand at the lingual side. Now let him say to the patient, "Pass the second finger of your right hand down till you feel my finger-nail, then press against your gum, and hold steadily." Now the operator may remove his right hand, and is at liberty to place the clamp. The aid of the patient judiciously applied is frequently as useful to the dentist, and more satisfactory to the patient, than the help of a third person. In the upper jaw, however, the patient cannot assist so easily. He will get his hand in some awkward position, which interferes with a clear view of the work. Having placed the dam, using both hands, the second finger of the left hand along the buccal side, that hand may be turned so that the palm is uppermost. In this position, and without moving the second finger, the forefinger may be passed into the mouth and made to take the place of the finger of the right hand, which has been resting against the palate. The dam is thus held with the fingers of one hand, the tooth appearing between. The clamp may then be easily adjusted with the other hand.

Ligatures.—This subject comes up naturally in the consideration of the dam, but I shall speak of it not only in this connection, but from various other stand-points. Indeed, the ligature is more valuable in other work than when used merely to force the dam into place and hold it there.

To tie silk around the neck of a tooth is generally so painful that it should be resorted to as infrequently as possible. Where it must be done, the application of a four per cent. solution of cocaine, freshly prepared, will do much to alleviate the suffering.

Ordinarily a ligature is not needed in placing the dam; certainly it is a rare case where more than two teeth need tying. In placing the dam over teeth, the edge of each hole naturally turns toward the incisive edges. If left so, leakage will result. By using a smooth flat burnisher, or other dull instrument, these edges may be pressed upward toward the gum till they become inverted. Immediately the contraction causes the edge to slide up under the free margin of the gum, so that unless the mouth is abnormally supplied with saliva the parts will be kept sufficiently dry for all practical purposes. This is a general principle, which, once noted, will cause the operator to use ligatures less and less frequently, as he becomes more and more dexterous in inverting the edges of the dam around the tooth-neck. When, because of the shape of the tooth or the superabundance of saliva, this cannot be relied on, a ligature is essential. It might seem unnecessary to tell how to tie a ligature, but there are numerous instances where a little more knowledge would benefit one who has but one way of doing this seemingly simple thing.

In the first place, how can the cord be tied tightest? and where is the best place to have the knot? It is easiest to tie a tight ligature around the central incisor, and the difficulty increases as the posterior teeth are approached. This is presuming that the knot is placed along the labial or buccal festoon. The reason is that while the hands are free at the median line, in the bicuspid or molar region the cheek prevents tension in one direction. There is a method, however, which overcomes this. Suppose that the cavity be in the approximal surface of a second bicuspid. In placing the ligature around the first bicuspid, pass it first between the two bicuspids, then around the first bicuspid so that the two ends protrude toward the buccal surface. Make a surgeon's double twist, and before drawing the knot tightly pass the proper end between the cuspid and bicuspid. The two ends may now be grasped and the knot tightened readily. because both hands have free play. The knot thus occurs between the cuspid and bicuspid. The second tie is made in the same way, except that a single twist is all that is necessary. Before cutting off the ends, pass the one on the buccal side down between the teeth, so that both protrude toward the palate. The ends, when cut off, are out of the way, and cannot stand up in the line of vision, as often occurs when the knot is made at the labial festoon. The ligature around the second bicuspid is made similarly, the knot occurring between that tooth and the molar.

While I advise placing the knot at the approximal side of a tooth, it should be occasionally along the labial festoon. We find anterior teeth, especially cuspids and lateral incisors, so shaped at the palatal festoon that it is very difficult to keep the dam from slipping down. If a ligature be used, and pressed up around the neck with an instrument, it is often found as soon as tied that it has passed above the dam, and nothing has been accomplished. This may be overcome by tying a good solid knot in the ligature first. When applied, this knot should be in line with the palatal sulcus. It gives a point of resistance for the instrument used to push it under the free margin of the gum, and carries the edge of the dam with it. When using this method, it is best to have the tying knot at the labial festoon, for being opposite the other it makes the ligature more secure. If it be desirable to have the loose ends out of sight, they may be carried between the teeth and cut off on the palatal side.

I now come to a description of ligaturing which may seem dry reading, because of the difficulty to describe knots in words. To him who has heard that teeth may be firmly united with flax threads, and has doubted it, I say, take a plaster cast and a piece of silk, then follow my directions by doing on the model what I describe, and the result will, I am sure, repay the trouble, and the reading will no longer be unprofitable. Unless this be done, even the assistance of illustrations will not render the description perfectly intelligible.

By many it has been claimed that we cannot fasten teeth permanently with ligatures. Certainly it cannot be done with wire, which either slips or breaks, and silk or flax it is said will loosen after a time. Again, the use of thread of any kind is considered uncleanly, because the brush will not remove the débris which accumulates between the ligature and the gum margin. This latter claim is due to the fact that most men tie a ligature *near the neck of the tooth*, if indeed it is not so placed as to actually impinge upon the gum. If such a ligature be worn for any considerable time, gingivitis will probably ensue. I will describe how to tie together a whole set of teeth, if there be need, so that they shall be held firmly, *the ligatures being about the middle of the teeth instead of at the necks*. Before doing so, I wish to touch on the use of ligatures in connection with the filling of teeth which are either loose or sore from wedging.

I shall first consider the last specified condition. Two teeth having been spread apart by wedging, are slightly loose, and sore to pressure. Because of some necessity, let us suppose that it is impossible to defer the operation. The teeth must be filled at once, yet the hand-pressure, or mallet, used for condensing the gold causes excruciating pain. Proceed as follows : Imagine that the cavities are at the median line, in the anterior approximal surfaces of the central incisors. The dam is made to include six teeth. A stout flax thread is waxed and passed between the lateral incisor and the cuspid on one side. A surgeon's double knot is made so that it occurs between the lateral and central incisors. This being firm, the two ends are passed around the central incisor, and the double twist drawn tightly along the anterior approximal surface ; the central is thus drawn tightly against the lateral. The knot being completed, the two ends are next carried back and tied along the posterior surface of either the lateral or the cuspid, should the lateral not be firm enough to give good support. The same proceeding on the other side of the mouth produces the result that both central incisors are bound firmly to their neighbors. To complete the stability a wooden wedge must be forced between the incisors, giving all the space possible, and acting as a keystone to the arch. Teeth thus held may be filled with gold when otherwise they could not be touched.

When from pyorrhea or other causes a single tooth is found to be quite loose, it is frequently advisable to unite it to its neighbor with a gold filling. It has been advised to apply the dam and then encase the teeth with oxyphosphate of zinc (some use plaster of Paris). This is sometimes best, especially with the anterior lower teeth. Where the tooth is either a molar or a bicuspid, I have usually been able to obtain satisfactory stability by the use of ligatures. In Fig. 49 are shown a first and second molar, with a continuous groove from one to the other, and ligatures binding the two together. Only two rows of ligatures show, and these would be insufficient ; but I have allowed the artist to draw but two, lest the figure be confusing. The dam being in position, the waxed flax is passed between the first molar and the bicuspid, and a double knot made between the molars, in the manner already described. Then the ends are carried back around the second molar and tightly tied along the posterior surface, the loose tooth, whichever it be, being thus drawn tightly against the firmer one. This is what I term the figure 8 ligature, and is to be followed by the figure O, which is accomplished by bringing the ends around and tying them along the anterior surface of the first molar, Fig. 49 a. Thus the two teeth are brought toward each other by different ways. In Fig. 49 the figure 8 ligature is seen nearest to the gum, and passing between the teeth, while the figure O is the higher one, seen to span the space. As has been already said, these two alone are not sufficient for the purpose, but a further stability is attained by repeating the process. If possible, the thread should be cut off at the outset long enough to serve for all the tying. If this be done, the next step would be to work backward once more, using the figure 8, and then forward again with the other. If the teeth are thus enwrapped six or eight times (that is, three or four of each), the loose tooth will not only be firmly held against the other, but will be supported so that the mallet does not cause pain.

FIG. 49.

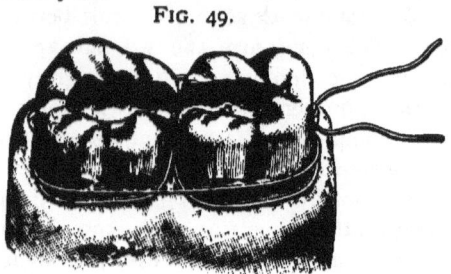

If the ligatures are well and neatly placed, one will lie above the other so that when completed they resemble tape.

I now come to the ligaturing of four or more teeth, whether for a temporary or permanent purpose. I may as well, perhaps, enumerate some of the circumstances wherein such tying becomes useful. Where teeth are badly loosened by pyorrhea, so that cure is either impossible or else must result only after a course of treatment covering a year or more, if they are to be united, it will probably be best

FIG. 49 a.

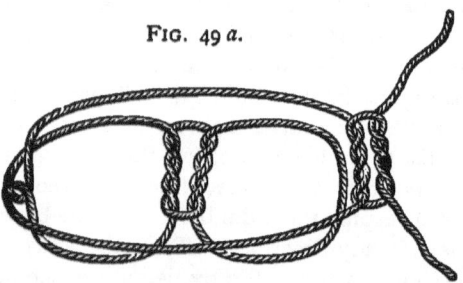

to accomplish this by continuous gold filling or some method other than resorting to ligating. Where the disease is in an incipient stage, or where the teeth are in a mouth which it would be wrong to disfigure by a show of gold, then the ligature must be depended upon. Again, the flax thread may be relied upon in cases of teeth loosened by fracture; for holding regulated teeth during twenty-four hours while a retaining plate is being made; for the support of one or more loosened teeth, elongated so that shortening by file or corundum wheel is necessary ; or for any purpose whatsoever where it is desired to hold for a period not exceeding six months, a number of teeth in a fixed position.

Let us imagine a case such as I was recently called upon to treat. A portly gentleman fell against a curb while crossing a street, and striking against his upper lip, loosened the six anterior teeth, breaking the corners from one or two and chipping the others. Whether he calls immediately after the accident, or foolishly waits three weeks (as my patient did), until inflammation has set in and one or two fistulæ have appeared, the first thing to be done is to make the teeth firm. In the instance cited, I accomplished this with flax so well that not only were the teeth made so firm that mastication became at once possible, and under treatment all were saved, but I was enabled to grind the jagged edges from the teeth without discomfort, although before the tying it was painful for him to touch the teeth with his tongue.

By the aid of Fig. 50 I will endeavor to show how this may be accomplished. In the figure are seen two rows of ligatures. The

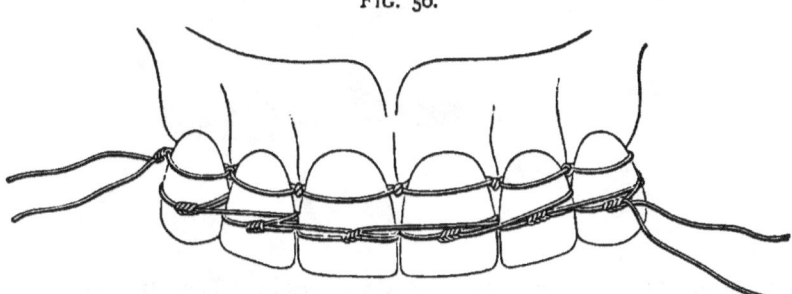

FIG. 50.

upper, or one nearest to the gum margin, is simply an extension of the figure 8 ligature, so that, as seen, it includes six teeth. One point must be mentioned in this connection. As already described, the double twist is made between the teeth and drawn tightly, one end of the thread being toward the palate and the other toward the lips. In the posterior teeth I have directed that the second half of the knot should be similarly made. In the incisor region the knot must be completed with both ends presenting at the labial aspect. By following this advice, the action of drawing the teeth together is attained while tightening the double twist between the teeth, but were the knot completed in the same manner it would make a bulge at that point which would prevent close contact when the next tooth was included. By having both ends at the labial aspect, the bulge of the knot occurs at the labial angle, where it cannot interfere. Thus the knot does not show in Fig. 49, while in Fig. 50 one is seen at every space. In the figure it may be observed that the ligature starting at the cuspid has been carried along, tying each tooth, till it finishes at the opposite cuspid, where the ends are seen. Once more I have left the work incomplete, lest the illustration should be confusing. In practice I should after reaching this second cuspid reverse the order of work and proceed backward again, tying in each tooth, till I ended where I had begun.

In Fig. 50 is seen a row of ligatures nearer the cutting-edges of the teeth. These are applied in a different manner, and this method I term "weaving." If the reader will consult Fig. 50 a, or else operate on a model as before suggested, this most valuable method will be made clear. In this case the tying is begun around the right cuspid, the knot being made at the center of the labial surface. There are now two loose ends. One is carried inward, passing between the cuspid and the lateral incisor, then outward between the lateral and central incisors. The other end is carried straight across the space, passed inward between the lateral and central incisors, and then out-

FIG. 50 a.

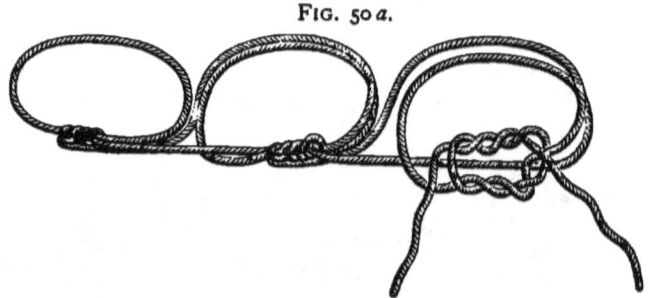

ward between the lateral incisor and the cuspid. Then the ends are tied once more into a knot, this time at the center of the labial surface of the lateral incisor. In a similar way the "weaving" is continued till the opposite cuspid is reached, when the ends are cut off short, no effort being made to return. In some instances I use both these methods on the same set of teeth, as, for example, in cases of fracture, but I must call attention to the fact that they are totally different in their action. The figure 8 simply draws the teeth together laterally, while the "weaving" will force outward any tooth which stands within the arch. This in many cases is desirable, while in others it is to be avoided. Where the four inferior incisors are loose, and the cuspids either absent or of such form that a ligature cannot be tied securely, if the weaving method were used about the four teeth the result would be that they would stand in a straight line, which would be mischievous. In such a case the arch must be braced with the figure 8 ligature first. If the cuspids cannot be utilized, then begin around a bicuspid and skip around the cuspid. On the other hand, it sometimes occurs that, though all the teeth are loose, one is much more so and stands within the arch. Here the weaving alone will usually not only be sufficient, but will force the irregular tooth into proper position.

It is noticed that in the figure both rows of ligatures are well away from the gum line. To accomplish this the dam must be placed, and just before tying in each tooth the labial and lingual surfaces should be smeared lightly with sandarac varnish. Ligatures thus treated, and tied by any of the methods described, flax thread being used, will not slip for several months and are clean. Of course, all thread should be waxed.

Clamps.—The usefulness of a clamp largely depends upon its grip upon the tooth to which it may be applied, and this in turn is proportionate to its adaptation and to its spring force. When not in use the jaws should approach each other so nearly that the distance between them is about one-half of the diameter of the tooth which it is intended to encircle, and the spring should be so strong that when stretched over such a tooth it will bind it firmly. A clamp for a bi-

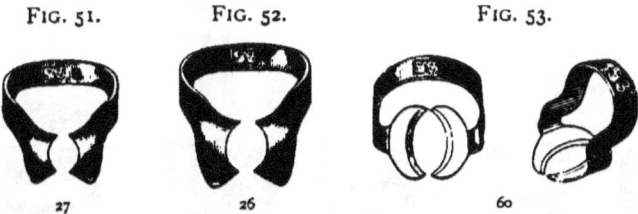

FIG. 51. FIG. 52. FIG. 53.
27 26 60

cuspid should have a spring so powerful that one could not endure the pressure were it placed upon the little finger, and one for a molar should cause discomfort if similarly placed. This is a fair test, and all which do not so pinch the finger would prove worthless in general practice.

If the practitioner can afford it, he should supply himself with a set of Delos Palmer clamps, because these are more accurate in their adaptation than any others which have been devised. Where the dentist is compelled to economize, he will be able to get along comfortably with just three clamps, excluding cervix clamps. These are bicuspid clamp No. 27, and molar clamps Nos. 26 and 60. Of these latter, No. 60 is adaptable to more cases than any one clamp that I have ever used, but it will sometimes occur that a longer arm to the bow would give a better view of the cavity, by stretching the rubber backward. Then No. 26 is preferable.

A clamp is not needed with any of the incisors or cuspids as a rule, except for cavities occurring along the labial festoon. For this position the ideal clamp, which will be universal in its application, is still to be invented. There are two, however, which, within their limitations, are most useful. One is Dr. W. W. Evans's form, Nos. 74 and 75. These are made either with or without a set screw. I believe that this screw has been practically useless to the majority of dentists, and yet it is really invaluable. The inventor says of it, "The office of the screw is to tighten the hold in cases where the clamp does not fit securely." While this may have been the object aimed at, it is a purpose not served for several reasons. In the first place, it not infrequently happens that the clamp is not pressed up sufficiently far to allow the screw to reach the tooth at all. Secondly, when it does touch the tooth it rests upon a sloping surface, so that if it is screwed down it displaces the clamp. Again, were it to reach a perfectly flat plane, at the very best part of the tooth for its action, a moment's thought will show that to screw it down would have a tendency to lift the jaws up from the tooth, and so loosen the clamp. Therefore as a screw it is useless, but as a projection above the surface it is most valuable. Supposing the tooth to be filled is a cuspid, the procedure

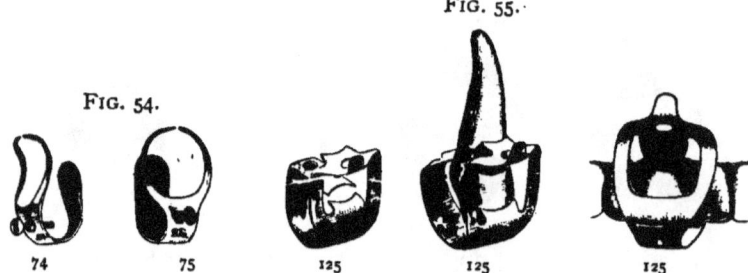

FIG. 54. FIG. 55.

74   75   125   125   125

would be to apply the dam over several teeth, and then tie ligatures about the necks of the lateral incisor and the bicuspid, leaving long ends. Next press the clamp high up on the tooth, and, holding it firmly, make a knot with the loose ends of the ligatures from both teeth, around the projecting head of the screw, so that when drawn tightly it tends to force the clamp upward. Completing the knot makes the clamp secure, and the operation may be finished without fear of slipping. Of course, for this purpose, a small hook would answer in place of the screw-head, but we may utilize the clamp as it is furnished to us. Therefore the form with the screw is preferable to that which is without it.

The peculiar shape of this clamp makes it specially useful in cuspids, the bulbous portion of the tooth at the palatal side near the gum offering an excellent point of resistance to the concaved jaw, while the two narrow projections above admirably fit about the labial surface and offer good view of the work. At the same time the clamp is limited in its usefulness by the fact that these projecting jaws frequently hide a portion of the cavity should it encroach upon the approximal surface. In favorable cases I have filled the portion along the festoon, and then have removed the clamp in order to reach that part which it covered. It is an exception, however, when this can be done without the ingress of moisture. Another class of cavities where this clamp cannot be employed is where there has been extreme recession of the gum about the neck of a long cuspid, followed by caries which reaches the receded gum-margin. The bow limits the distance to which the clamp may be made to reach.

For teeth other than the cuspids the How cervix clamp, No. 125, is the best, known to me. The inventor has recognized the fact that a clamp for these cases should have three bearings upon the tooth, because of the fact that the palatal jaw will not bite on a line with the labial. Being bowed over the cutting-edge of the tooth, it has the same limitation as the Evans.

I will dismiss the subject of clamps by saying that, excepting festoon cavities, the clamp should preferably be placed upon a tooth other than the one to be filled ; also, in applying a clamp care should be taken not to press too hard against the sensitive gum, and especially not to bite the gum itself, as is often done on the palatal side.

Leakage.—When we recall that the dam is placed over teeth for the purpose of keeping out moisture, it is plain that when leakage occurs our object has been frustrated either wholly or in part. Much has been written to account for the failure of gold fillings along the gingival border. I think ninety per cent. of these failures may be attributed to leakage during the operation. Moisture is easily dammed up in a dry mouth, but in one abundantly supplied with saliva the greatest skill is required to keep a cavity absolutely dry,—so dry that at no time after placing the dam can even a trace of moisture be observed.

It will profit us to consider somewhat carefully this subject of leakage, especially in relation to failure at the gingival margin.

Let us suppose that the cavity is in the approximal surface of a central incisor, and so closely approaches the gum that a ligature is needed to bring the gingival border into view. Before placing the dam the parts are all wet ; so after the dam has been placed this moisture must be wiped away, and this is done. The dentist then proceeds to prepare the cavity, when, after using a chip-blower, he is surprised to find that some of the dust still adheres near the upper border. This shows that there was moisture. He concludes that he had not wiped it thoroughly dry, and does so now. He uses the engine-bur a second time, and when the dust has been blown away there is still the same evidence of moisture. Having thoroughly dried the part before, he is now satisfied that there is a leak. Examination shows none, or rather none that is discernible to the eye. Therefore, as he can find no leak, he is satisfied that there is none. To be quite sure, however, the cavity now being ready, he uses the warm-air syringe, and proceeds with his filling, placing a handsome piece of work, beautifully polished. In a couple of years he finds this with decay about the gingival border, and perhaps writes a learned essay upon the instability of fillings at this point.

Now, what are the facts? When the dam was first placed and the parts dried, all seemed well. As soon as the engine was used, moisture appeared, recurring a second time. Lastly, the spunk and hot air made all dry once more, and the eye, watching, could detect no moisture. Therefore the mind reasoned, "No leak." Nevertheless there was a leak, or else how did moisture twice recur? The nature of this leakage will be explained in a moment, but just here I must elucidate this enigma and show its relation to the subsequent failure of the filling; otherwise it will be said, "A leak which did not interfere with the operation of filling was of no consequence."

When the dam was first placed and the parts dried, all moisture seemed to have been excluded. This simply meant that the dam was effectual *as long as the parts were in a state of rest*. Then the engine-bur was used, and moisture appeared. This is accounted for by the statement that *whenever a patient is hurt, fluids are discharged into the mouth*, whether by nervous or by muscular action is immaterial. The fact is the important thing, and it is a fact. Thus after the use of the hot-air blast, the parts being at rest, all seeming dry the operator proceeded. In doing so he made pressure upon the tooth, which caused pain, thereby inducing a further ingress of moisture. The pain of packing the gold being less than that of using the bur, the fluid accumulates more slowly. Thus it does not interfere with the placing of gold along the extent of the border. About this time, however, moisture, unseen, slowly dampens the gold at this point, so that a *tiny* pellet would not cohere. But the operator is not using tiny pellets, but pieces of some size, *which he first attaches to the unwet gold within the cavity, and then mallets down over the border*. Thus it may happen that each piece of gold coheres along a portion of its extent, so keeping its place, *while there is no cohesion whatever along the gingival border*. Such a filling may be polished and burnished so as to present a fine appearance, but there is a weak spot, and it is along the gingival border, where the leak allowed a slight percolation of moisture.

I must now explain the existence of a leak which is undiscernible by the eye, and yet allows the entrance of moisture. It most often occurs because of the fact that the hole punched for the passage of the tooth is made too large. In a very wet mouth, if the dam is slipped over a tooth readily, the chances are that one of these invisible leaks will occur. Therefore, in such mouths the holes should be cut so small that they just escape tearing when stretched over the teeth. In this manner the tooth is hugged so tightly that the contractile action of the rubber compels it to reach as small a diameter as possible, which is found up under the gum-margin, because of the conical shape of roots. Of course in a dry mouth any tendency to leakage caused by a loosely-binding dam may be, and usually is, overcome by the ligature. In the case above described, where there was barely room for the silk above the border of the cavity, the ligature itself operated against the discovery of the cause of leakage, while it in no way prevented the excessive moisture from creeping under it, and immediately upon the gold.

Another cause of leakage is seen in Fig. 56, which shows the dam over two teeth whose crowns are omitted in order that the points of leakage, *a*, may be more readily seen. Here the fault has not been

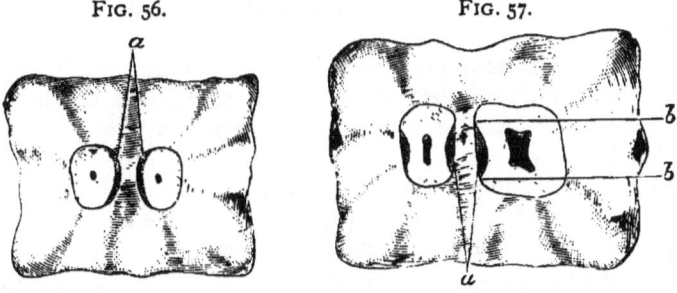

FIG. 56. FIG. 57.

that the holes were made too large, but insufficient space was left between them, and the stretching necessary to place the dam over two teeth has produced a space next to each, as pictured. Of course this may be avoided by proper spacing. In the next diagram, however, Fig. 57, is shown a similar difficulty arising from a different cause. Here a section through the teeth near the gum-line discloses a distinct concavity along the approximal surfaces. The dam stretched over such a tooth must follow a straight line from the points *b*, *b*, so that a space is unavoidably left. The question arises, Can the dam be placed here so as to avoid leakage? Moisture cannot be kept out by the dam alone, but, the condition being recognized, a roll of cotton dipped in sandarac varnish should be packed along the leak; then a ligature should be knotted so as to present three or four knots, which should be placed over the cotton and the silk then firmly tied around the neck of the tooth. In a few moments all leakage ceases.

Another occasional source of leakage is where, in placing the dam so as to include a molar, the edge is not forced between the teeth at the posterior border of the molar. The clamp is placed, and all seems well, but in a wet mouth moisture will inevitably creep in. Indeed, there are many instances where leakage will occur about a clamp which will cease if the clamp be removed and a ligature tied about the tooth before replacing it. Again, it may happen that there is a space along a palatal or labial surface of a molar similar to that shown in Fig. 57, as occurring at the approximal side. This is seldom found in normal conditions, but often occurs where there has been recession which reaches the bifurcation. In these cases the roll of cotton dipped in sandarac should be laid over the part, and held in position by the jaws of the clamp, which are pressed against it.

FIG. 58.

A slight tear in the dam, occurring at a time when the filling is so far advanced that the dam cannot be renewed, may sometimes be remedied by dipping a piece of sponge in sandarac, and then forcing it into the hole. The same thing may be accomplished by making a button of wood, as shown in Fig. 58, which may be whittled in a moment from a piece of orange-wood.

The best course is not to tear the dam, and much trouble may be avoided in this direction by doing all the preparation of the cavity before placing the dam, except in cases of sensitive teeth, or those in which the pulp is suspected to be nearly exposed; then the cavities should be prepared dry. In using a sand-paper disk, if the edge is allowed to run over a piece of soap it will not cut the dam, and is less likely to become entangled in the folds of rubber. The dam is frequently cut with the sharp points of clamps, so that care should be exercised in that direction.

The Napkin.—The napkin I believe I use more than many, and differently from some. The small napkin made into a roll I find much use for in the lower jaw, but seldom upon the upper. It should be about three inches square and without fringe. In all ordinary cases it may be folded into a roll and applied about the teeth in the form of a horseshoe—the ends forward. It is readily held in position with the mouth-mirror. Occasionally it is of advantage to secure it more firmly, in which instance it may be caught against the labial and lingual sides of the tooth with a clamp. A special clamp has been devised for this, but any good molar clamp will serve well enough. In cases where a lower sixth- or twelfth-year molar, recently erupted, is too short to permit the use of a clamp, and yet the condition is such that the operator deems it advisable to fill with gold, the napkin may be best held in place with the napkin-holder, which is a metal horseshoe with an arm which slides in a ratchet having a piece to fit under the chin. This may be arranged to hold a napkin securely for half an hour or longer, at the same time forcing the tongue out of the way.

Sometimes the only necessity for using a napkin is the fact that the patient is oversensitive and easily nauseated by the dam. If there is danger of moisture arising between the teeth, a piece of rubber-dam just large enough to cover the two teeth and immediate vicinity may be placed without inconvenience to the patient, and then the napkin in combination with this will keep all perfectly dry for a short time.

Where I use a napkin in the upper jaw, I simply fold a large napkin once, so as to present a doubled edge, and pass this back behind the tooth to be filled, holding it with the mouth-mirror. This is less annoying to the patient than any other method, and will keep a crown cavity dry long enough to insert gold. I have filled the posterior approximal surfaces of wisdom-teeth in this way, producing permanent gold fillings. This, however, is the most difficult place for using a napkin, owing to the fact that the muscles offer tremendous resistance.

Chloro-Percha and other Devices for Controlling Moisture.—Chloro-percha may be made to serve a very useful purpose at times. A case may arise where it is impossible to apply a clamp to a festoon cavity, which is nevertheless so small that it could be quickly filled if only kept dry; a napkin will not serve, for, though it might be made to dam up the saliva, there is often a weeping of mucus from the membrane overhanging the cavity, and this is sufficient to prevent the insertion of a cohesive gold filling. If the gum be wiped as dry as possible, and then smeared with a thin coating of chloro-percha, the cavity may be filled successfully by a skillful, rapid operator.

There is a device of which I have made frequent use, but which I believe is now difficult to obtain. It consisted of a sort of clamp, which, however, was not needed, and concaved disks of pipe-clay. It is well known that a fresh clay pipe will stick to the lips. These disks, made of the same material, could be placed over the orifice of the duct of Steno, and would adhere, shutting off that supply of saliva for an hour or two. In many cases they prove most useful.

Many use Japanese bibulous paper for damming up saliva. Its advantage over the napkin is that like the clay disk it adheres to the mucous membrane, but it absorbs moisture more rapidly. For short operations, however, it is frequently a good plan to make a stiff roll of this paper and stuff it between the cheek and alveolar process, thus covering the opening of the duct.

There are little rolls of cotton which are now sold for this same purpose, and are useful.

Wedges vs. *Separators.*—I have the boldness to assert that steel separators have worked more harm than good to the profession. I have no doubt that in the hands of the most skillful, and used with discriminating judgment, they have saved much time for this limited class; but in the practice of the many the result has been too often the filling of teeth with insufficient space. The man does not live who can fill as perfectly in a narrow crevice as he could in an approximal surface fully exposed by the loss of a tooth. It follows logically that as much space should be obtained as is possible without permanent injury.

There are several reasons why a perfect filling cannot be made where the space is very slight, but the chief one is that gold should be built out so far beyond the true surface, and over all borders, that the final finishing will cut down to a perfectly condensed part of the gold. For this reason, where teeth are forced apart with steel separators, it often happens that the fillings when finished show pitting, or at best will not take a mirror-like polish, which should be made on all, but more especially upon those in approximal surfaces.

There are two kinds of cavities which require considerable space. These are very small, and very large cavities. To fill a tiny cavity through a narrow crevice is to court failure, while it is simply impossible to reach into all parts of large cavities, in long bicuspids, without considerable room. In the latter, all that part which is most inaccessible is improperly filled, but being also out of sight is not seen by the patient, and not looked at too closely by the dentist, who thinks he cannot afford the time required for properly separating teeth.

If it is important to fill a tooth at all, it is important to fill it well, and to fill it well the cavity must be accessible. This can only be accomplished by separating as widely as is consistent with safety. The cases where teeth can be forced as far apart (with the patient's consent) by using a separator, as by the old-fashioned system of wedging, are so few, that to buy a separator seems to me to throw away money, for I refuse to insert gold with insufficient room for thorough work.

Where the patient is easily hurt, if there is no hurry, the teeth may be painlessly forced apart with tape. The patient should be supplied with this and directed to change it, increasing the number of thicknesses from day to day until the space is adequate.

The rubber wedge is the most positive, and at the same time the most painful, usually making the teeth so sore to the touch that they must be allowed to rest for several days. This necessitates the removal of the wedge and the insertion of a piece of gutta-percha. This is placed between the teeth cold, trimmed just thick enough to slightly wedge when pressed in. With a warm burnisher the surplus is then removed on each side, and the soreness will subside very rapidly, the teeth meanwhile being kept apart. The main objection to rubber is the tendency to slide up against or under the gum. This may usually be prevented by allowing a bit of the wedge to extend beyond the cutting-edges of the teeth. Sometimes it is well to apply the dam, smear the teeth lightly with sandarac, and then place the wedge, after which it will not be apt to slip. The dam can be removed by first cutting the septum of rubber between the teeth.

Where it can be used, I like the wooden wedge better than any other. To apply it, first cut a thin wedge, soap it, and slip it between the teeth next to the gum. Next trim the wedge proper with a long taper, soap it, and force between the teeth, driving it into place with gentle taps. After cutting off the ends, withdraw the wedge which was first placed, thus relieving the pressure against the gum.

Whenever there is sufficient space above the gingival margin of the cavity to permit it, a wedge should be forced between the teeth during the operation of filling. This not only supports the teeth, lessening the shock from the mallet blows, but it protects the dam when the disk is used. This wedge should be made thin and shaped as seen in Fig. 59. The grooves at each side make it more firm, and prevent it from riding up. As the wooden wedge is not always entirely satisfactory, owing to the fact that if proper grooves be cut it sometimes becomes difficult or impossible to force it into place, I have devised a steel wedge which serves admirably in most cases. This is shown in Fig. 60. It is similar in general shape to the wooden wedge, but a V-shaped cut forms the sides into two arms which can be sprung together, and so forced into place, where it is retained by the strength of the spring. It will be noticed that this has a tendency to further separate the teeth; this is slight, and no disadvantage.

FIG. 59.   FIG. 60.

The Uses and Dangers of Matrices.—If separators have worked evil, matrices have proved even a more disastrous delusion. It seems so easy to prove that a tooth which has been encircled by a matrix cannot be properly filled with gold, that it is astonishing that so many really skillful men use them.

The argument is this: What is the principle upon which a matrix is used? It is made to supply a lost wall, and thus to produce a cavity which is practically similar to one which has all walls standing, which latter is admittedly the least difficult to fill. This the matrix does, and since the cavity which it simplifies is usually most difficult, at a casual glance it would appear that the matrix is a most valuable instrument. The fallacy lies in this, that while (perhaps) it renders the actual filling-process more simple, *it forms the gold so that it becomes impossible to properly polish it.* To explain this the illustration (Fig. 61) will serve. The matrix is seen drawn tightly around the bicuspid. The teeth have not been wedged, since no space is needed save for the matrix, which usually can be crowded between the teeth in their normal position: indeed, it is usually preferred that the teeth should be close together. It seems to me that the logic which I am now about to use is simply unanswerable. The tooth being filled, and the matrix removed, it is certainly true that *any polishing done with sand-paper or file must remove a portion of the tooth, however little, and thus produce a permanent space.* I have refilled teeth because this had been done. But it may be contended that it is not necessary to polish between the teeth when a matrix has been used, since the gold packed against the matrix must be already smooth. Let us consider this. I have contended that all fillings, and especially those in approximal surfaces, should be made as smooth as a mirror. To produce such a surface by simply packing gold, it follows that the surface against which the gold is packed should be as smooth as glass. This is not true of any matrix, but can this be done? I presume that few men would claim more skill than that which was possessed by Dr. Marshall H. Webb. Several years ago I had the honor of taking part in a friendly contest with him and several others, to test various methods of inserting gold. We used

FIG. 61.   FIG. 62.

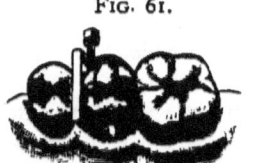 

different kinds of mallets, and filled glass tubes set in wood, these being held in place with cement. Here we had practically, *glass matrices.* After the fillings were made they were tested in various ways. To determine the adaptation to the walls (glass) they were placed in an aniline dye. *All of them leaked badly.* They were then removed from the tubes and examined with a magnifying glass. *All of them were badly pitted, the pits showing plainly even to the naked eye.* Thus it follows that several gentlemen, Dr. Webb among the number, failed to make gold adapt itself to a smooth surface. Therefore they could not pack gold against the smooth surface of a matrix, so that upon the removal of the matrix the gold would be polished. Who will claim to do in such a case what Dr. Webb could not? It may be asked, If these fillings leaked, why do not fillings in teeth leak? The answer is that in the tooth the gold is packed against a *rough* surface, which by offering resistance makes it possible to obtain a sufficiently accurate adaptation.

It will be claimed that a matrix may be made of a material which will yield, so that under the force of the mallet the gold will bulge beyond the proper contour, and may then be polished. This is true of the upper two-thirds, but is not true of the gingival margin. *There is but one way to produce a perfect approximal filling; that is, to ob-*

tain sufficient space so that the gold may be made to extend over all margins, and bulge beyond the normal contour, so that when all is trimmed away a perfect surface is produced at all parts, and of such contour that the teeth returning to normal position will knuckle. This is impossible with a matrix.

There is no objection to using a matrix around a tooth when the cavity is next to a space made by the loss of a tooth (Fig. 62), for here the filling may be polished, the slight loss of tooth-substance being of no consequence. The greatest use for a matrix is where a large part of a tooth must be contoured with amalgam, for this is a material which, because of its plasticity, would be most difficult to use without the support which the matrix gives. Moreover, it is a material which can be polished with very little loss of substance. Where it is possible, the matrix should be allowed to remain in place twenty-four hours.

To sum up, a matrix is invaluable when amalgam is to be inserted, and disadvantageous where gold is to be employed.

CHAPTER III.

The Uses of Various Filling-Materials—Methods of Manipulation —Materials of Minor Value—Gutta-percha—Oxychloride of Zinc—Oxyphosphate of Zinc—Amalgams—Copper Amalgam—Gold.

An endless variety of materials has been suggested for replacing the portion of a tooth which has been lost by caries. But few have proven of great value, and no one material has been found to satisfy the demands of all cases. It is, however, no more necessary for the dentist to rely upon one filling-material than it is for the physician to use one drug with which to control all the diseases of mankind. We can intelligently care for all conditions, provided we have at least one method which will be useful in each given condition.

There is but one material which is best adapted for the filling of a given cavity at a specified time. Let us use that specially indicated material on every occasion, even if we should lose by the transaction. Let us refuse to use amalgam where gold should be employed, and equally let us fill with amalgam if best, even though the patient would pay more for gold. Few would deny the wickedness of placing a number of amalgam fillings in the incisor region, yet if gold can be used to best advantage it is equally wrong to improperly fill bicuspids and molars with some other material.

If one is to follow this rule of never filling except with that which is best adapted to the circumstances, it becomes necessary for him to be able to determine for all cases what material, or combination of materials, is best fitted to the occasion. Therefore, in speaking of the various materials, I shall endeavor to state where each is positively indicated, and leave the reader to discriminate for himself in less obvious positions.

The main reliance of dentists is upon gutta-percha, the plastics, and gold, but there are other materials which have been used, and of them I will make cursory mention.

Materials of Minor Value.—Lead has been used in children's teeth, and occasionally in well-defined cup-shaped cavities in adult mouths. Thin sheets of the metal are cut up into narrow strips and rolled into convenient sizes. They are retained in place by a process of wedging. This at once limits the cavities in which lead may be used to those having well-defined, strong surrounding walls. If omitted from the office, I believe it would never be missed.*

Tin has been used similarly to lead, with no special advantage. It has also been made into foil and used as soft or non-cohesive fillings are made. Before the introduction of amalgam, it occupied the same place which that material too often does now, for the filling of teeth *where a high fee could not be charged*. Except for this purpose, *which is an unprofessional one*, it has little value. Latterly, in combination with gold, it has been claimed to have special therapeutic effects not possessed by either metal alone. It is doubtful if this claim can be proven to be well founded. I will say more of it when considering gold.

Robinson's felt is a shredded metallic composition, which has a place in the office. It is claimed for it that gold will cohere with it, so that a cavity may be half filled with it and completed with gold. This is a delusion against which the inexperienced must be warned. It is true that in favorable cavities two-thirds of the filling may be made of the Robinson's felt, and gold may be used for completion, but there will be no cohesion. The union is purely a mechanical one, and only of sufficient tenacity to make manipulation easy. The felt must not extend so near to the surface that the remainder of the cavity would not of itself retain the gold. The felt must not be packed too tightly, but left sufficiently loose so that with a heavy serrated instrument the gold may be packed and forced into the felt. This will be readily accomplished, and the cavity may be completed with cohesive gold. The advantage of this method is that large cavities may be rapidly filled, and as well as if all gold were used. The cases where this is exclusively indicated are large molars from which the pulps have been removed, tremendous cavities needing to be filled. In many such cases it is preferable

---

*I am not here considering the filling of root-canals.

to use the felt rather than to have too large a mass of oxyphosphate underlying the gold, or to force the patient to submit to the exhaustive process of packing gold alone. It is claimed that this material also makes a good filling for temporary teeth, but in my judgment other things serve better.

Porcelain fillings, so called, are pieces of tooth-porcelain specially baked to fit a given cavity, and then retained by a cement. Hence it is obvious that the durability of such an operation depends upon the cement, which cannot be counted permanent, though more nearly so here than when used alone. The only excuse for using porcelain is the hope of matching the tooth in color. This is only accomplished by accident. The color of a given mixture will depend upon the degree of heat at which fusion occurs, and as this cannot be definitely measured, it becomes impossible to bake porcelain specially and produce a desired shade. This alone is enough to make the method disappointing, if not worthless. Porcelain inlays are different, in that the filling is made from porcelain already baked, so that color and shade may be selected to match, if a sufficient variety be at hand. Up to the present time inlays have been successfully made for only a small number of cavities. If a method should be devised by which the porcelain may be accurately ground to fit an irregularly-shaped cavity, there will be many cases where the inlay will be indicated: notably upon the labial surfaces of superior incisors.

Vitreous, or glass fillings, are made from a kind of glass, ground to a fine powder, which fuses readily over an alcohol flame. It is furnished in an assortment of shades, which it is claimed are reliable; that is, that the powder from a given shade will always produce a given color. If this should prove to be true, this method may be useful occasionally. Of all kinds of fillings, however, which depend upon cement for retention within a cavity, I am inclined to believe that only temporary results will be obtained, and the practitioner should be cautious in making promises to the patient in whose mouth he attempts what is as yet in an experimental stage.

Gutta-Percha.—No dentist can afford to be without this most valuable material. Useful in many ways, it is perhaps most useful for temporary purposes. In the form known as "temporary stopping" it is immensely valuable. This is a combination mainly of gutta-percha and wax. Of those obtainable at the depots I prefer Gilbert's, because it has just the right proportions of the two materials. Those which have more wax are not as durable, and become too soft and sticky when warmed. The temporary stopping is furnished in two colors, presumably that the white may be used in conspicuous positions. There is a much better advantage to be taken of the two colors. The pink should be used only to cover arsenical dressings, or in such teeth where the condition demands that attention should be given to the tooth at the next sitting. The white becomes useful for teeth whose root-canals have been filled, or any other condition where it is not absolutely essential that the particular tooth should be operated upon immediately, to the exclusion of other possibly urgent cases. Thus, when a patient presents, a glance in the mouth tells much. A temporary filling, white in color, the dentist may set aside for the time ; but a pink one acts as a signal, which would mean that delay would be dangerous.

A temporary purpose, which is served by gutta-percha plain, better than when in combination with wax, is where a patient presents with a number of dangerous cavities ; teeth in which a pulp-exposure may occur at any time. It is impossible to fill them all at one sitting with permanent materials. It is very wise, however, to cleanse them all of decay, and fill with gutta-percha. Thus all is made safe at once, and the permanent fillings may be placed at leisure.

As a permanent filling gutta-percha may frequently be depended upon. Either the white or the pink may be used, but my experience has been that the pink is more durable ; therefore in inconspicuous places it is to be preferred. Much of the reported failure of gutta-percha as a permanent filling may be referred to faulty manipulation, or injudicious choice of the cavity in which to place it.

So far as manipulation is concerned, the common practice of heating the material in the flame is ruinous to all hope of permanency. It should be heated on a porcelain disk held over the lamp, or preferably over warm water on a glass tray. A special apparatus of this kind is purchasable. In placing gutta-percha in a cavity, should the cavity be a large one, it may be packed piece by piece, thus insuring adaptation to the walls, until two-thirds of the cavity is filled. Then a single piece large enough to complete it should be used. In smaller cavities a single piece should be chosen large enough to slightly more than fill the cavity. After the filling has cooled and is hardened, the surplus should be trimmed off with a thin smooth burnisher, or spatula, slightly warmed, care being used not to drag the material away from the walls. In a few instances, as up under the gum margin, I have used to advantage a tape dipped in chloroform.

Places where gutta-percha is positively indicated are extremely sensitive cavities along buccal, palatal, or lingual surfaces, especially where they extend wholly or in part below the gum margin, in molar teeth. Here the pink variety is to be preferred. Where teeth have been worn by ill-fitting clasps, so that cavities filled with leathery decay have been formed, which when cleansed leave exposed hypersensitive dentine, they are more likely to be cured with gutta-percha than with any other material. It is better to be obliged to renew the fillings periodically than to risk death to the pulp by using a metal. Gutta-percha is not a non-conductor, but it is a poorer conductor than the metals, and what is more important, all the tissues of the mouth, whether hard or soft, are singularly tolerant of it. Occasionally a patient will present with a deep cavity in a most inaccessible approximal surface. Excavation makes it doubtful whether the pulp is nearly approached, or whether hypersensitive dentine alone is the cause of the pain. A very wise course is to temporize by inserting gutta-percha. Too much examination may expose a pulp in a small cavity, thus necessitating a tremendous sacrifice of tooth-substance for its proper removal, whereas under a gutta-percha filling this class of teeth is frequently troublesome no longer.

Gutta-percha is often used as a capping over pulps which are nearly approached. Where this is done, and the filling completed with gold, common sense will indicate the advisability of covering the gutta-percha with a resistant body of oxyphosphate before attempting to pack gold upon it. Many, however, would not think this essential where amalgam is to be the final filling. It must be remembered that gutta-percha is

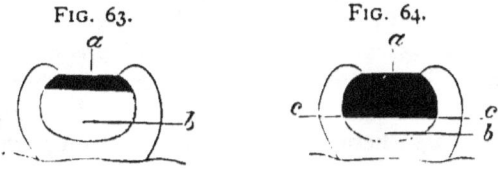

FIG. 63. FIG. 64.

slightly elastic. Therefore where it underlies amalgam or other metallic filling it must not be so placed that the superincumbent filling rests wholly upon it. In Fig. 63 is shown a cavity filled improperly. The amalgam, a, rests upon the gutta-percha, b, in such a manner that pressure during mastication has a tendency to compress the latter, allowing the amalgam to be driven downward away from the walls, thus producing leakage. In Fig. 64 the arrangement is better, for the amalgam, a, is now supported by the walls at the points c, c, so that movement under pressure is rendered impossible. Thus it becomes important not to insert too much gutta-percha under another filling in the vain hope of obtaining better insulation.

A solution of gutta-percha in chloroform, commonly known as chloro-percha, is useful for lining cavities before inserting a filling of oxychloride, thus preventing the pain from irritation by the chloride of zinc. It is also valuable, and should always be used as a coating for oxychloride or oxyphosphate fillings where the latter are meant to serve permanently.

Oxychloride of zinc, as a filling-material, is much less used than it was before the introduction of the oxyphosphates. As a permanent filling it is doubtful whether there are any conditions in which it is to be preferred to oxyphosphate. By some, however, it is earnestly claimed that if a plastic of this form must be used where erosions have occurred, the oxychloride will serve better than the oxyphosphate. This is a statement which may be true, but as yet it has not been proven. It may be true, but before it can be known to be true it must be explained why it should be so. This will not be possible until the etiology of erosion is less shrouded in mystery than it is at present.

One claim advanced is that the chloride of zinc, being a powerful germicide, is advantageous. Before we may depend upon oxychloride fillings for the reason that chloride of zinc is germicidal, it must be shown that *this action continues even after union with the oxide into a solidified mass.* Again, it must be proven that there is a relation between bacteria and erosion, and the special bacterium must be isolated. Then it must be shown that the oxychloride filling placed in a cavity will cause a resistance to the erosive action which may result from the presence of the bacterium. Meanwhile it is hazardous to depend upon oxychloride as a permanent filling in cavities near the gum-line. Innumerable instances have been reported and observed where in such positions a seemingly well-made filling has become wholly disintegrated, not only allowing decay around it, but apparently itself decaying. As this special case, for which this material has been advocated, does not really demand its use, it would perhaps be better to depend upon oxyphosphate where a plastic for a permanent purpose must be used.

Another advantage which this material is said to possess is the power to excite the tooth-pulp toward the production of secondary dentine. The idea is that in deep cavities where the pulp has been nearly approached, but not fully exposed, if oxychloride be placed as a capping, its therapeutic action will result in a deposition of secondary dentine by the pulp, so that the distance between the pulp itself and the bottom of the cavity is materially increased. Were this the case it would be most fortuitous, for no capping can be so good as tooth-bone itself. That a pulp has the power thus to protect itself is undoubtedly admitted, but I think I am safe in saying that it is a physiological action, rather than pathological. Under the slow wasting away occasioned by erosion, this new barrier is thrown out so that the pulp is constantly protected from the invasion. Erosive action has seldom been known to expose a living pulp, but we frequently see erosions which have passed beyond the original limits of the pulp-chamber, and into the new dentine which the pulp had formed as a protection. The question arises, whether an oxychloride filling can induce the same result? The difficulty of proof lies in this, that where it did accomplish this, there would be little temptation to remove the filling or the tooth for examination. Thus its failures can be more accurately counted than its successes. Another fact to be considered is, that just as in the case of erosions, the physiological action of the pulp would have the same tendency toward self-protection against the advance of ordinary caries. Indeed, where the carious action happens to be slow we not infrequently find it aborted by the resistant quality of the new dentine formed, so that we get what has been termed "arrested decay." It follows from this that any filling whose action is sufficiently negative not to interfere with the ordinary healthy action would seem to produce the desired result, merely because it does not prevent it. It is probable that where the oxychloride succeeds, it acts just as other materials do, by not creating an irritation.

That an oxychloride filling does possess some therapeutic qualities seems indicated by the fact that it may usually be depended upon to lessen the sensitiveness of dentine if left in a cavity a month or more. This of course may occur under one of oxyphosphate simply because the tooth itself has altered, but as it has in my observation occurred more frequently with oxychloride I deem it safe to say that an oxychloride filling has an obtunding effect.

The oxychloride being obtainable whiter than oxyphosphates, is preferable for lining discolored teeth where the walls are thin. Occasionally it may happen that it is desired to line a wall, and yet there may be a need to utilize as much of the cavity as possible for the proper retention of the filling. The thinnest imaginable layer of oxychloride may be effected by smearing the wall with the fluid and then blowing the powder against it with a chip-blower. It should be noted that this is apt to destroy the rubber bulb, because the oxide of zinc taken within the bulb acts deleteriously upon rubber, causing it to dry and crack. This may be prevented by careful washing after use.

Oxyphosphate of zinc is invaluable both for permanent and for temporary purposes. As much depends upon the manner of manipulating the material, I will explain how best to mix it. According to the method of mixing, the resulting mass will be either quick or slow setting, and either dense or crumbly.

First, then, as to its setting. The best mixing slab is the side of a flat glass bottle, which should be provided with a tightly-fitting rubber stopper. In the winter months, with the temperature of the room at 70° F., the bottle has no special advantage unless it is desired to make the material set either very slowly or very rapidly. If the bottle be filled with ice-water, the setting will be slow in proportion to the cold, and *vice versa*, if filled with hot water, it will be rapid in proportion to the heat. This is more useful in summer, when all have experienced much inconvenience because of the rapid setting of oxyphosphate, so that in warm weather a bottle of ice-water is very useful.

In mixing the material, the method must depend upon the use to which it is to be put. If a permanent filling is desired, put the powder and liquid upon the slab separately. With a clean, smooth spatula stir a little of the powder into the liquid until it is incorporated, producing a thin but well-mixed material. Add more and more of the powder until a thick, smooth cream is produced. This, of course, is still too thin to be used as a filling, *but no more powder must be added.* With the spatula continue to work the mass, when it will soon be observed that it begins to thicken and ball up on the slab. It now has a sticky quality, and may be used if it is necessary to depend upon adhesion to the cavity wall. If this is not the case, further working will produce a mass which may be taken between the fingers and worked as we do gutta-percha. Made into a roll, and cut into pellets with the sharp edge of the spatula, it is in a most convenient form to be packed into a cavity.

If oxyphosphate is to be used for setting a crown, it should be similarly mixed, and used at the creamy stage without further kneading. For *temporary* purposes, where it is desirable that the filling may be readily removed at a subsequent sitting, as, for example, where it is employed to cover an arsenical dressing in a shallow cavity, after the creamy stage, instead of kneading to produce thickening, continue to add the powder until the consistency is suitable for use. A mass thus made may seem the same as the other, but really it is quite different. There is an excess of the powder, and after hardening it can be more readily crumbled away with a sharp excavator. Thus there is a vast difference between the preparation for permanent or for temporary purposes. May not some of the failures recorded against the material be more properly attributable to its faulty manipulation?

It has been advised that during the mixing some germicide should be worked into the filling. In this way oil of cloves and other medicaments can be stirred in with no apparent harm to the material. Moreover, the odor of the disinfectant will be noticeable very long after, but whether there be any advantage in the process remains to be demonstrated.

Oxyphosphate as a filling-material has many uses, the most important of which is as a mass to be interposed between a metal filling and the tooth itself. It must be remembered that save in rare cases gold or amalgam furnishes no support to frail walls. In fact, the metallic filling depends largely upon the strength of the cavity walls for its permanency. The oxyphosphate, because of its adhesion, does support frail walls, and therefore is peremptorily required in all such cases. Unlike gutta-percha, there need be no limitation as to the quantity used, save that the cavity above it must be of a shape which shall be retentive for the metallic filling which is to cover it. In these cases it is preferable to place a portion of the metallic filling while the oxyphosphate is still plastic, for thus the upper filling is practically cemented into place. This process will be described later.

As a permanent filling oxyphosphate may be used in cavities where the more conductive properties of metal would prove injurious. It also should be employed in conspicuous positions, as, for example, large corners or labial festoon cavities in the incisor region, in the mouths of actors and actresses, ministers, singers, lecturers, and public speakers generally.

Relative Values of Amalgams.

Amalgam is one of the most valuable filling-materials at our disposal. At the same time it is the most abused. It is more frequently chosen because of its cheapness than because of its special adaptation to a given case. The result is that it is not properly manipulated, and is too often allowed to remain without special polishing. Much of the disrepute attached to amalgam is probably due to faulty filling methods, rather than to any bad quality inherent in the material itself.

One of the charges against amalgam is that it shrinks, thus producing imperfect borders, which invite decay. How much this may be due to using the material mixed of an improper consistency, is worth considering. I believe that the shrinkage of amalgam is reduced to a minimum by proper mixing. Another thing worthy of note seems to be the curious fact that this shrinkage does not appear to involve the entire mass, for if it did would not the filling become loosened within the cavity? This I have never seen. The discrepancy occurs at the surface, so that it would appear that all we need is a method of manipulation which will overcome this tendency to shrink along the border.

If it is necessary to use the dam for the insertion of gold, it is equally needful where amalgam has been decided upon. I do not mean that the dam must always be employed, for I have explained that it is not always a necessity even with gold, for which it is more often used than with other materials. It is more exact, then, to say that the cavity must be kept as dry as possible, whether by using the dam, or by dependence upon the napkin. I think that amalgam inserted without regard to moisture is more likely to become blackened than when caution is used to prevent the wetting of the material during its insertion.

In mixing the alloy with the mercury I prefer adding the mercury, a little at a time, until a plastic mass is produced, rather than to use too great a quantity of mercury at the outset, and then depend upon expressing it. If mercury, squeezed from amalgam mixings, be preserved in a bottle, and then examined, it will be found that it contains a considerable proportion of metals which it has absorbed from the alloy. Thus it would seem that it has brought away with it a portion of that metal contained in the alloy for which it has the greatest affinity. The result must be an alteration in the proportion of metals contained in the alloy. Thus, if a given alloy is used for making an amalgam, and one mixing is made with just the proper quantity of mercury, while a second is prepared with an excess which is removed by pressure, it would seem natural to expect that the two would not act similarly even in the same mouth. This would explain, perhaps, the oft-claimed unreliability of an amalgam, where it will fail in a tooth upon one side of a mouth, and succeed admirably upon the other. It is not the amalgam that is unreliable, but the dentist, who has not a definite method.

As it would be a tedious process to weigh out the alloy and mercury for each mixing, we may resort to some less accurate way, provided that it accomplishes the result with reasonable reliability. If the attempt be made to mix alloys with mercury in the hand, it will usually result in an excess of the latter. We may better depend upon the pestle and mortar, preferably of small size. Place in the mortar the amount of alloy desired, and add a little mercury; stir vigorously, and the mercury will take up all the alloy with which it can unite. Add more, and more, stirring between each addition, until the mass is soaked with mercury, but perhaps granular. If it now be turned into the palm of the hand and manipulated vigorously, the friction produced will aid in the amalgamation, and a mass quite plastic can be produced. This plasticity is not the result of an excess of mercury, for we have seen that while cold, in the mortar, it did not reach such a consistency; it has been attained by heat rather, and there is no need to squeeze out any mercury, though, in order to be sure that there is no excess, it is as well to squeeze with the finger and thumb. That amount of pressure will not force out any mercury if too much has not been used, and if this method of mixing be followed it will rarely occur that any excess will be found.

The amalgam is now ready for use. To place within a cavity as large a mass as can be crowded in, would be as incorrect as to follow the same rule when using gold. Choose a piece which can be dropped into the cavity easily, and which under pressure will become packed *without fracturing*. The homogeneity is thus preserved. With a bit of bibulous paper rolled into a small ball and held in the foil-carriers, the amalgam is next pressed, and smeared into all the crooks and corners of the cavity. Usually this first portion of amalgam may be treated so that it will form a veneer covering all the walls. This insures adaptation. Piece after piece is to be placed in this way, and if the mass was properly mixed no excess will come to the surface.

while under the action of the ball of paper a hard filling is produced at once. In large cavities a burnisher may be used over the paper, to exert more force than can be attained with the foil-carriers.

When the cavity is filled, the next step is to follow the process which I say will prevent shrinkage. Indeed, it will do more, *for it will preserve a handsome " white" color for years.* Alloys which contain a large proportion of gold have not proven successful. Nevertheless it is gold upon which we must depend for this most desirable improvement in amalgam fillings. Take gold foil No. 3 (No. 4 may be used, but the thinner grade is preferable), and tear it into small pieces without folding or rolling. When the tooth has been filled with the amalgam, place upon it a piece of this thin foil. Then with a *warm, smooth burnisher* gently stroke the gold till the mercury in the amalgam absorbs it. This is not the old scheme of using tin to extract mercury from an amalgam filling. In that method the tin does not enter the filling, but acts in removing the mercury, as a sponge does in sopping up water from the floor. *The gold, on the contrary, must be forced into the filling, so that it changes the surface into an amalgam of which gold is a component part.* Gold is to be manipulated in this way, upon and into the amalgam, until it will no longer be absorbed. This can be continued until the surface of the filling is quite hard, and even golden in color. This color will not be retained after the crystallization, but the filling when polished at a subsequent sitting will be found of fine texture, and will receive a mirror-like finish.

Where a large contour is to be made, involving say two or three cusps of a molar, an excess of amalgam is to be used, and then when the gold is applied a density will soon be reached which will allow the operator to carve up cusps, by removing the excess of material and forming sulci, after which more gold is to be burnished into the filling until the surface is apparently set. A large filling of this character may be dismissed with little fear of seeing it crushed into shapelessness at a subsequent visit. Moreover, it is much easier to trim up a contour while the material is thus semi-plastic than to be obliged to defer this for another day. It thus becomes apparent that if the best results of amalgam are to be reached by burnishing gold into its surface, we cannot hope for the most perfect fillings when we use a matrix, and do not obtain a separation of the teeth in advance of the filling. This is only one more argument against the matrix, which I have said is useful only with amalgam, whereas now we see that even with this material it has its disadvantages, since it prevents the burnishing of gold along the approximal surface.

There is another method of manipulating amalgam to which I will allude, since it is little known, though I do not care to take the responsibility of recommending it, for the reason that though I have used it with seeming success, I have not done so for a length of time sufficient to make me ready to indorse it.

When filling with amalgam, we often have an excess of the material which we throw away as useless. If, instead of this, these surplus pieces are preserved, they may be used thus : where a large filling is to be placed, with the usual danger of fracture, during mastication, before perfect crystallization has occurred, this may be prevented by using the bits of amalgam left over from previous mixings. These of course have crystallized. Take one in the carriers, and hold it in the alcohol flame until softened by heat. Quickly place it upon the filling, and with a *hot* burnisher force it into the mass, which will immediately become hardened. With a little practice this may be done very satisfactorily. Whether it will prove durable or not, I am not prepared to state absolutely. It is to be remembered that it is *superficial* shrinkage which we are desirous of preventing, and may this not accomplish it, *since we use for the surface a bit of amalgam which has already crystallized out of the mouth ?* Besides, this is very similar to the advocated method of using copper amalgam, except that the latter becomes plastic under the application of heat, and retains this consistency for a considerable time, whereas the ordinary amalgam returns to a crystallized form almost immediately, so that rapidity of manipulation as well as skill is requisite.

Before passing from this subject of the rapid hardening of amalgams, I must note a fact which I have never seen published. Often, after mixing a large mass, a longer time becomes requisite for packing the filling into the cavity than had been anticipated. Thus a portion of the material, not yet used, begins to set, so that it is difficult to manipulate. I have resorted to a heated burnisher to restore its plasticity, with the invariable result that while I have succeeded in incorporating the amalgam with the filling, the surface has been immediately hardened. Thus I think it safe to assert that heat, while packing amalgams, hastens their setting.

I now approach the important subject of *where* to use amalgams. I am not a believer in the material to such an extent that I would advocate the too common practice of deciding that any cavity which is in an inconspicuous location may be as well filled with amalgam as with gold. Too many patients have been so educated by contact with dentists, that they will say, "It will not be seen, doctor; therefore fill it with amalgam." This is an error, though the converse is true,—*i.e.*, "It *will* be seen ; therefore *do not* fill it with amalgam," though even to this last rule there must be an occasional exception, as far back as the bicuspid region. I think it would be a wise rule which would rigidly exclude amalgam from incisors and cuspids, while the bicuspids should be included as often as possible.

Cavities in the crowns of molars, though out of sight, should be filled with gold, the rule being relaxed only as the cavity becomes larger. If a cavity be small, and therefore of that class which may be safely filled with anything and be preserved as long as the duration of the material, *it is the very place for gold*, because gold is the most durable and reliable of all materials. A phosphate or a gutta-percha filling will stop caries, but these must be considered temporary, and amalgam preferred to either. For a similar reason gold must be chosen instead of amalgam. As we come to consider cavities of larger size the rule becomes less binding, till at last the point is reached where it becomes a decidedly difficult question to determine which material must have preference. One way of reaching a decision is to consider that the seemingly frail walls will still support a gold filling, because of the fact that the pulp is still living. In a similar cavity, where the pulp has been removed, it might be safer to depend upon amalgam. After this point has been passed, and the cavities presented for consideration are of great size, amalgam should be the first choice, and gold used only where everything combines to make a perfect operation possible. Thus, if the patient is of good health, with strong nervous stamina, and willing to endure fatigue, it becomes possible to place a large gold filling, with the hope that the last third may be as well condensed as the first. Too often operator and patient undertake that which, because of the tediousness, becomes so great a tax, that in the end the work is hastily finished, and thus improperly completed. Either the gold is not packed with sufficient force to produce a resistant density of surface, or there is not sufficient material used to allow of proper contour or occlusion. Again, given favorable circumstances as to health and strength of patient, the welfare of the tooth itself must be considered ; that is, whether the placing of a gold filling would become a hazardous operation. Where a recent attack of pericementitis has been aborted, it might be attended with unfortunate results to work upon the tooth for a period long enough to fill it with gold.

Thus, as I have said, as the cavity increases in size, the demand for amalgam is correlatively increased, though a gold filling properly made is always preferable.

An important place for amalgam is where the cavity-border is below the gum-line. No filling will succeed here unless the finished surface along the gingival border can be made scrupulously smooth. There are many such cavities which should be filled with amalgam rather than with gold, solely for the reason that the amalgam may be made smooth more readily than gold. Many such may be filled with gold by a dentist dexterous in placing the dam, and yet the filling after completion may defy the efforts of even a skillful operator to properly polish along the gingival border. Approximal trimmers will do much, but at best they leave a filed surface. Would we deliver a gold plate which showed file-marks? If not, why should we leave a filling in such a condition? It must be remembered that the cavity-border up under the gum-line may be accessible while the cavity is empty, but the placing of a properly contoured filling must make it more than ever inaccessible, so that the gingival border may become utterly concealed. An amalgam filling may be made flush at this important point while it is yet plastic, and the utmost care should be used to make it so at the time of filling, for if a gold filling is difficult to polish, an amalgam filling, after hardening, is even more so.

It is claimed by some that it is good practice to build with amalgam from one approximal cavity across into another in the adjacent tooth. This seems to be of doubtful merit, and rather to be condemned than advocated. However that may be, it is generally admitted that a filling in an approximal cavity must not jut out and press upon the gum-tissue. Moreover, care must be taken not to force particles of amalgam under the gum, which by being left may bring about a condition of ulceration. Where the dam cannot be placed, I have found a means of accomplishing this very nicely. Take a bit of bibulous paper and make a tight rope of it, thick enough to just pass between the teeth. This rope is to be laid against the gum and pressed up beyond the cavity-margin, which can be done more often than would seem possible to those who have not tried it. This effectually prevents any material from passing beyond the gum-line, and when drawn out after the tooth is filled leaves a well-defined space, wherein the burnishers may be used for properly shaping the filling along the gingival margin. Nevertheless, in spite of the efficacy of this, it is usually advisable to thoroughly syringe the parts with warm water, using a syringe which will throw a stream with considerable force, placing the point at the gum-line, so that the water passes between the teeth and washes out any débris.

In spite of the fact that I have here advised the use of amalgam in cases where the gingival margin is above the gum-line, yet I would distinctly deprecate the use of this material for patching gold fillings which have failed in that locality. I think it would be better to remove the leaky filling entirely before refilling, for certainly I have seen numerous successful cases where the amalgam had been placed along the gingival border first, and a completion of the filling made with gold subsequently. Let me not be misunderstood, therefore. It is not the use of amalgam in connection with gold that I deprecate, but only its use as a *patch* near the gum-margin. This brings us to the consideration of amalgam in combination with other materials. It has been successfully used with gold in two ways. The most widely known is where the amalgam is placed along that part of a cavity which, being below the gum-line, is impossible to keep dry. This is filled with amalgam until the dam can be placed so that the rest of the cavity may be dried thoroughly. The amalgam is allowed to set, and at a subsequent sitting a portion of it removed, leaving as little within the cavity as possible, without risking the slipping of the dam. The gold is then placed, being partly anchored into undercuts made in the amalgam. This is often advisable. Especial cases arise where a large portion of an anterior approximal surface of a bicuspid has been lost by caries. Restoration with amalgam alone would be unsightly, and with gold alone impossible. The amalgam may be placed along the gingival fourth or fifth of the cavity, and covered with gutta-percha, till the next sitting, when it will be found that the dam may be used and the gold packed into the remainder of the cavity. When all is done, nothing shows but the gold.

The second method is to pack the gold upon the fresh amalgam. To accomplish this a matrix is required, and the plastic golds are preferable to form the contact with the amalgam. This kind of gold may be readily packed against the amalgam, and will soon assert itself by overcoming the efforts at amalgamation to such an extent that the filling may be completed with cohesive foil at the same sitting. A filling of this nature made in a glass tube an inch long, one-half of each metal, has been so successfully made that the resulting rod when removed from the glass showed a perfect union, with great strength, between the gold and the amalgam. The objection to this, in my mind, is that the matrix must be used.

It will frequently occur that a cavity is so shallow and of such poor shape for retention that it is questionable whether an amalgam filling would be retained. Here we may frequently depend upon a method which is exceedingly useful with gold also. Mix oxyphosphate to a sticky consistency, and smear some upon the walls of the cavity, immediately adding some amalgam. Allow the phosphate to set thoroughly, and the amalgam is practically cemented to place. The filling is then continued by adding more amalgam, and will be more likely to remain in place because of the underlying phosphate. Care must be taken, before adding the second batch, that no phosphate be left overlapping any of the margins, as that would leave a weak point.

In a similar way, in deep crown cavities, we may depend upon the phosphate, not only as a protection to the nearly approached pulp but as an additional retentive precaution, provided we pack the amalgam before the phosphate sets, instead of waiting till afterward, as is the usual custom.

The mixing of an amalgam with a phosphate I have never tried, and can see no advantage in. The mixing of the alloy with the phosphate, however, is totally different, and in many cases a practice deserving of much praise. The dry phosphate powder is mixed with the fine filings of an alloy, and afterward used in connection with the liquid, exactly as plain oxyphosphate fillings are made. The presence of the metal interferes with the mixing only a little, and the usual putty-like consistency may be attained. A filling of this kind is to be placed and allowed to set two or three days. It may then be polished by using a smooth stone in the engine.

I have seen these fillings advocated in all positions, but I consider them more useful in masticating surfaces than elsewhere. The disadvantage of any cement filling is that it slowly dissolves out. By adding the alloy we have an ordinary phosphate filling which holds a number of particles of metal in close contiguity. As with any other phosphate, the material itself gradually wears away. As the metal does not dissolve, it is plain that the surface becomes roughened, as the metal filings become more and more exposed. For this reason it is not wise to use the method in approximal surfaces, since the roughened filling will invite decay. But in the masticating surface a totally different result obtains. As fast as the metal filings are exposed, they are flattened down by attrition, till at length the whole surface is metallic, and can be burnished just as an amalgam filling may be. After this, the phosphate being thoroughly protected by this metallic veneer, there is no further waste.

These fillings are especially good in temporary teeth, though I sometimes use them in adult mouths. When I do, it is because of their low conductive power. Where a cavity is very sensitive, and it is deemed advisable to place a filling temporarily until such time as when, the sensitiveness having been controlled, gold may be inserted, this combination of an alloy mixed with a phosphate is excellent.

Copper amalgam has been so much in vogue during the last few years that I can scarcely pass this subject without alluding to it. I have only a few facts to mention, and will leave deductions to the reader.

When copper amalgam was introduced into this country as a substitute for the ordinary alloys, its advocates claimed that aside from not having many disadvantages attributable to ordinary amalgams, it possessed a *therapeutic* quality, in that no carious action could recur in its vicinity. *Practice has not substantiated this claim. Copper amalgam fillings, from the hands of practitioners known to me clinically as expert operators, have come under my observation leaking badly.*

When a copper amalgam filling becomes thoroughly blackened, is then well polished, and finally assumes a bright ebony-like surface, it seems to be the most admirable tooth-saver known. This is a success. When, on the contrary, it does not become black, but remains a dull, lusterless gray, it will be found readily removable with an excavator, and will waste away as surely as a phosphate filling. This is a failure.

Were it possible to determine in what mouths, and under what conditions, these two results could be prognosticated, we would be enabled to give copper amalgam a prominent place in our cabinets. But when it is known what many have observed, first, that fillings which appear successful for months will suddenly deteriorate into a condition similar to that described as a failure; second, that fillings may be successful, and unsuccessful, in the same mouth at the same time; and third, that two fillings have been made at the same sitting, from the same material, placed in the same tooth, both in the masticating surface, and that one of these has proven successful while the other has utterly failed, we are compelled to admit that at present copper amalgam is utterly unreliable. It has been suggested that the failures are attributable to the manner of manufacturing the material. This may be so. But until the true secret of manufacture shall have been discovered, copper amalgam cannot be recommended.

Gold as a Filling-Material.

Gold is pre-eminently the best material with which to fill teeth. It has the great disadvantage of being a good conductor of heat, and it unfortunately requires considerable time for its proper insertion, in addition to which it lacks some of the good qualities of other materials; but nevertheless it can be stated positively that the main reliance for the salvation of teeth which have decayed must be upon gold.

It then becomes of the gravest importance for a dentist to thoroughly understand this material and its manipulation, and after acquiring the requisite skill, to be conscientious in his application thereof.

It is furnished to us mainly of two kinds, cohesive and non-cohesive, and in three forms, viz, foil, including cylinders and blocks, rolled or heavy foil, and the various kinds of plastic gold.

In using gold, the dividing line is where we decide to depend upon non-cohesive or cohesive gold. Non-cohesive gold was used by the earlier dentists exclusively, because the dependence upon the cohesive property of the metal is comparatively recent. Since the adoption of cohesive gold, the older methods have fallen into such disuse that it is safe to say that only a small percentage of dentists still follow them, while a very large proportion of the younger men have never even essayed them. The question then arises, "Is there any advantage in non-cohesive gold which is sufficiently important to make it deserve a place in office practice?" If there is, I am ignorant of it. In making this statement, I am aware that I will be criticised, but my position is just this: In my own experience I have never discovered any special use for non-cohesive gold. I have never been able to appreciate the advantage of having my gold so that the different particles would not cohere under pressure, nor have I been able to see the disadvantage of having them cohere. I have frequently met prominent men who have admitted using non-cohesive gold. I have invariably asked these men why they ever prefer it to cohesive gold, and the replies have always been either evasive or unsatisfactory. The best answer ever offered to me is that non-cohesive gold may be used where a cohesive gold filling would be an impossibility, as for example, in the buccal surface of a molar in the lower jaw, where the cavity extends so far below the gum-margin that the dam cannot be employed. It is claimed that a good gold filling can be made in such a situation if non-cohesive gold be used. I believe that this is possible, but I also believe that it is possible only in the hands of a limited few who have attained to extreme skill in the manipulation of the non-cohesive golds. I think that as a practice for the many a large percentage of failures would be the result, while the same men could produce admirable and sufficiently permanent results with other material than gold.

A use which I was taught to make of non-cohesive gold is to place it against the walls of cavities, finishing with cohesive foil ; the idea being that non-cohesive gold can be better made to adapt itself to the walls, and that thus a more nearly water-tight filling will be produced. This I think must be condemned, especially where by so doing we would occupy, with non-cohesive gold, that part of the cavity upon whose shape we depend for the retention of the whole mass of the filling. While there will be some cohesion between the underlying layer and the superincumbent portion which is made with cohesive gold, I am certain that a stronger filling can be made where cohesive gold is used throughout. I have removed so many fillings, taking out first a solid piece, and then picking out the remainder in crumbs, that I believe the practice has been relied upon by many. My removal of such fillings has usually been because of discolored margins, showing where caries had crept in and was undermining the fillings.

My opinion, then, of non-cohesive gold is that in practice we have so little, if any, need for it, that the dentist may discard it almost *in toto*. In the colleges, however, its management should be taught, for the student who learns to fill with non-cohesive foil will have acquired a degree of manual dexterity which will quickly make him expert in the use of cohesive gold. Too many have neglected to learn in this way, and *the cohesiveness of gold has enabled many inexpert men to practice dentistry, palming off upon patients fillings which may look bright and smooth to the inexperienced eye, but which in a few months become rough, in which condition they are a menace to the safety of the teeth.*

Of the manipulation of cohesive gold much could be written. It is my purpose here only to point out the more important facts, and if possible to draw attention to a few points of faulty working, upon which I think many failures depend. I like to have all three kinds of cohesive gold in the office. A cavity may be filled with either, exclusively, and in many smaller cavities it is wise to do this. There are many other cases where the employment of two and sometimes of three kinds of gold is expedient.

Of the plastic golds, I have had the most experience with Watts's crystal. This is a form of gold much praised by many, and much condemned by others. Its strongest advocates undoubtedly use it too frequently, and place too much dependence upon it, while those who condemn it utterly are simply ignorant of its virtues. They have probably endeavored to use it as they do other plastics, forgetting that it is a metal. Its chief virtues are : first, its great cohesiveness, no gold having more, if any has as much ; second, the small amount of force necessary for condensing it ; and third, the fact that it will remain where it is placed, having no tendency to ball up, and roll about in a cavity.

Its first advantage, cohesiveness, makes it especially useful in many instances, of which I will enumerate a few. It has occurred to many that while inserting a large filling, leakage has made some portion of the surface of the partly-made filling non-cohesive. The effort to add another pellet of foil is vain. If the surface be touched with alcohol and then dried with hot air, crystal gold will adhere, *provided it is used in small pieces.* I have even removed the dam, where a large gold filling was but half finished, and continued the operation at a subsequent sitting, the gold meanwhile having been protected by gutta-percha, though of course this did not keep it dry. Such a procedure should not be adopted unless the remainder of the cavity still presents a retentive shape. Thus the adhesion of the crystal gold becomes more a matter of convenience than of necessity in such cases. Where a filling becomes wet before completion, it being in an exposed position, as the corner of a central incisor, if it be dried with alcohol and hot air, as before advised, and then a clean rose bur be used to freshen the whole surface of the metal, crystal gold may be made to unite firmly with it, and the filling may then be safely completed.

If, while filling, a bit of the edge of a cavity crumbles away by accident, or when finishing it is seen that there is a slight imperfection along the border, crystal gold is invaluable. The tiniest burs may be used to cut a groove or pit, and this may be more perfectly filled with crystal gold than any other, because of its cohesiveness, and more particularly because it can be torn into smaller bits without losing its pliability.

A third use is where a portion of a cavity which is very inaccessible requires the use of the mouth-mirror. Such a place might occur where, in an approximal cavity in an incisor, the palatal depredation has been great, and an examination of this palatal portion after the filling is nearly completed shows that it is not full enough. Crystal gold can be built on here more readily than any other.

The second good quality, its plasticity, makes crystal gold especially useful over nearly-approached pulps. It is frequently instructive for a dentist to have his own teeth filled. He then discovers, much to his surprise perhaps, that certain methods produce pain, which, as the dentist, he may have thought to be painless, or where patients have complained, he may have attributed the cause erroneously. I have myself been astonished to find that the mallet will cause excruciating pain over a particular spot; pain, too, of a character which showed that the pulp itself was disturbed. It is my opinion that where a pulp is nearly approached, that point of the dentine which is thinnest, becoming more resilient, yields slightly under the force of the mallet, thus producing pressure, which shocks the pulp. Therefore, when a patient complains of a distinct pain from the mallet at a specified point, and at that point only, or more than elsewhere, I now attribute the disturbance to the above-described cause, and continue the filling with crystal gold, using hand-pressure until, having built a sufficient thickness of metal over the thin portion of dentine, I find that the pulp no longer responds painfully to the mallet-stroke. For similar reasons small cavities may best be filled with crystal gold, using hand-pressure, whenever the tooth is sore from wedging. Still, except in the smallest cavities, I should prefer to make the extreme outer surface of heavy foil, as will be described.

Another most important usefulness of crystal gold is that its plasticity allows the operator to work with less danger of fracture along weak walls or in deep undercuts. A more resistant form of gold, acting as a wedge in such an undercut, may crack or even break the enamel. I have seen a whole corner forced off with a corkscrew plugger, which was being used to drive a pellet ahead of it. I think it could have been well filled and preserved by the use of crystal gold. This accident occurred at a clinic, and caused the operator much chagrin, but from an educational standpoint it furnished a better object-lesson than if it had not happened.

The third quality, that of not balling up, brings me to a description of the right and of the wrong way to use crystal gold. While it is true that it will not ball up if properly used, it is also true that it will so act in unskillful hands. The great danger with this material is the temptation to use too large pieces. The inexperienced may think that, being plastic, it may be crowded into a large cavity and then condensed afterward. I have seen this method advocated at a college clinic by one of the instructors. It is a mistake. In the majority of such cases, when the mass is condensed, it will be found that it can be moved to and fro within the cavity. *Whenever this occurs, with any kind of gold, the only correct procedure is to remove the whole and begin again.* To attempt to hold it in place, until wedged by what is placed above it, is to deceive one's self and defraud the patient. The mass may be thus fastened so that it will not move, but the space between it and the walls, *which must have existed since motion was possible,* will not have been eradicated. I cannot use language too positive in recommending small pieces of this or any other kind of gold. *I never place any bit of gold within a cavity which cannot pass the orifice without touching the edges.* I frequently use pieces even smaller than would be thus indicated. The rule is as binding with a plastic gold as with any other.

Another method of manipulation which I think is a grave error is to use a foot-plugger with crystal gold. A foot-plugger has a distinct usefulness, but it is an instrument which produces many poor fillings, simply because used in wrong places. I use small points for all kinds of gold, and equally so for crystal. Suppose that a mass of crystal gold be placed against the filling within the cavity, just as the gingival margin is to be covered. I should condense it with small points, thus uniting it perfectly with the gold already placed, and gradually approaching the borders. Then I should exchange for a foot-plugger, and go over the surface, flattening out hills, and especially perfecting the impaction against the border, finally chiseling off with this same plugger all the excess which projected beyond and over the edge. To sum up, then, I should depend upon crystal gold mainly for covering the walls of cavities, and for such other places as I have indicated. I should not use it much for contour work, or allow it to reach broad masticating surfaces. Care should be used to see that it is not partly condensed before it is inserted in the cavity. It should therefore never be held in the fingers. A good method is to make a gold wire loop, or staple, and with this pin the cake of gold to the wooden box in which it is sold. Small bits may be torn off with the foil-carrier as needed, without danger of compressing the mass.

Passing now to foil, I have to consider briefly the various forms in which it is supplied. In the first place, there is the condition of the surface. We find this either bright as though burnished, or rough, appearing as if "frosted." This style of gold has various names, depending upon the maker. I do not remember any special advantage that has been claimed for it, and it is probably manufactured to supply a demand created by a habit, rather than by reason. On the contrary, my own experience with it would condemn it. It is less cohesive than the plain, and this is seen to be natural upon a moment's consideration. The adherence of one thing to another may be produced at any time by effecting absolute contact. Glass, which we would not count as possessing the property of adhesiveness, will develop the property if two smooth pieces be wet and laid one upon the other. There is nothing sticky about the water, and its action is simply in completing the contact. Therefore *smooth* gold should be more cohesive than *rough*. I have been told that this frosted gold is more easily worked, that it is softer. That is admitting that it is less cohesive, since the words "soft" and "non-cohesive" have become synonymous when applied to gold. Moreover, this very softness is a disadvantage, since it renders the material crumbly, another property of non-cohesive gold fillings. If cohesive gold is to be used, it should be as cohesive as it can be obtained, and while it should be *pliable*, we need *toughness* in preference to *softness*.

Secondly, as to the form in which to use cohesive gold. Of course the packing of a continuous rope into a cavity is not to be thought of. An advantage with non-cohesive gold, it becomes a mistake with cohesive foil. The whole idea of such a procedure is to produce a filling made of as few pieces as possible. This is no longer needed when we can obtain a cohesion between the particles. It is therefore probable that the axiom should be reversed, so as to read, "Use as many pieces as possible" rather than "as few," as formerly. This, however, would be an extreme teaching, since it would mean the expenditure of too much time over an operation. It is, nevertheless, a sign-board pointing in the right direction.

We may obtain square blocks, and rolls cut into short lengths. To me, both of these are the lazy man's adoptions. He asks the manufacturer to do for him that which he should do for himself. As I do not like patent medicines, of which a stated dose will cure an unspecified individual, so neither do I fancy gold prepared to fit unseen cavities. How does it work? If we obtain an assortment, as we probably would, we begin instinctively by using up the smaller pieces. As we find the stock decreasing we are compelled to use larger and larger pieces, till at length we reach a condition where having a small cavity before us, and only large blocks in the gold drawer, we either improperly fill with too large pieces or else we stop to cut up the blocks, causing more or less compression, and generally poorly-shaped pellets.

I prefer, therefore, to procure gold foil in books, and roll and cut it for each case immediately before inserting it into the cavity. I have never found any way of keeping gold in the office, superior to leaving it between the paper leaves of the books. Foil in books may be obtained "trimmed" or "untrimmed." The latter sells at a lower rate, and is not only as good, but really is the same thing as the trimmed. This is a legitimate economy therefore, though it is a rare thing to find a strictly economical dentist who is at the same time a capable one. Dentistry is not a good field for economy. The patient should be given the best of everything, and the best is usually the highest priced. As good a way of cutting gold as any is to tear from the book two or three leaves, with the foil between, and with sharp shears cut both gold and paper. This avoids the unpleasant accident which frequently happens when gold is cut unprotected, where we find that it sticks to the edges of the scissors. These strips should then be loosely rolled into ropes, and cut into pellets suitable to the case for which they are prepared. To illustrate how far I carry this principle, I need only say that I have cut a strip as narrow as one-tenth of the width of the sheet, and have then made the pellets no longer than their width. I further state that this can be done so delicately that the pellets are proportionately as pliable—that is to say as loosely rolled—as where I used a third of a sheet, which is the maximum width to be resorted to. With tiny pellets of this character I fill tiny cavities; those made with the finest of rose burs, and which retain the filling because the bur is dipped first in one direction and then in an opposite. A filling of this kind may be smaller by far than the head of a pin, and yet have required a dozen or more pellets. I have more confidence in a filling so made, than I would have in one made with a single large pellet, however dexterously manipulated, and however bright it may be made to appear. I have seen a clinician fill teeth in this latter way, as a demonstration of "rapid work," and I have been impressed with the idea that we should make *rapidity* subservient to *perfection*.

There is an extreme danger which should be cautiously avoided in using cohesive gold pellets. I think it is usually preferable to fill undercuts with crystal, or some other of the plastic golds. Where this is not done, the foil may be used, but care must be observed in the direction now to be indicated. An undercut usually produces a weakening of a neighboring wall. If a pellet be taken which is much too large for the undercut, and is doubled on itself as it is forced into place, the instrument being between the folds, the whole acts as a wedge, producing a strain against that wall. This may fracture it so that it breaks away, or it may crack it so slightly that the accident may not be noted. Then, when decay creeps in, the usual sermon may be preached: "The tooth has decayed around my filling, but the filling is all right." It is well to be sure that the plugger chosen can be driven to the bottom of an undercut without danger, and then the pellet should be small enough to enter the space without doubling, though large enough to jam slightly; when so placed, it may be driven home with the plugger, which in this case is not *between* the gold, but *behind* it. The action is to compress the gold toward the bottom of the cavity, and not laterally between the walls and the sides of the instrument. This caution to use small pieces should be observed until the undercut is entirely filled. Another thing is that perfect cohesion is gained by this, and is not obtained if any other plan be adopted. How often have we all removed *some other man's filling*, taking it away in a solid mass but leaving gold in the bottom of the undercut? What does that mean? It means that the undercuts were presumably made, to assist in the retention of the filling, but because of the manner of packing the gold, no such result accrued. The explanation lies in this : Gold loses its cohesive property in proportion as it is condensed. Thus having partly filled an undercut, the gold placed therein remaining in position because it is wedged in, we have before us a small surface of gold presented, and an undercut lessened in depth. If the next pellet does not cohere, we would have been as well off with a shallower undercut. If a pellet which is too large be now used, the pressure of the plugger begins to condense it before it is made to touch the gold in the bottom of the groove, the result being that there is less cohesion, if indeed there is any. The principle, then, is to begin with a piece just large enough to be wedged in without lateral pressure, and to add next a similar piece, or one even smaller, so that it will readily reach the first, *and cohere with it before any great pressure has been exerted*. In other words, it should be condensed after it has cohered, instead of before. This one principle observed, will make approximal fillings permanent, which would fail where it is neglected.

For similar reasons, if a pellet by any accident has become slightly compressed, it should be placed in the waste-box. It should under no circumstances be put with the good gold, and finally used in some large cavity. That is an economy which is reprehensible, the more so because it cannot be detected. I saw this done once in a contour filling, by a friend for whom I was malleting. I subsequently saw the same filling, otherwise beautiful, with about one-quarter of its corner broken off. I thought at the time that I understood the accident, and I think so still. The filling had a flaw in it, and the flaw was caused by that bit of partly condensed, and, therefore, imperfectly cohesive, gold.

This compression of a pellet before it is brought into contact with the surface of the already packed gold brings us to another point of great importance. In the approximal surfaces of unusually long bicuspids we may be compelled to work with insufficient space. I mean space insufficient for the placing of a pellet of usual size without compression. The cavity may be one in which we would be authorized to use fairly large pellets ; it may be one in which to use very small pellets would mean to compel the patient to submit to too long an operation. Yet we may find that every pellet, as it is pressed to place either between the teeth or through the possible orifice in the cutting-surface of the tooth, becomes partly compressed. What must be done ? After showing that such pellets do not produce good results, it remains only to say that in such circumstances they should not be used at all. The entire filling must be made with rolled gold, otherwise known as heavy foil. This brings me to as good a place as any for explaining what I think the relative values of the three forms of gold. The use of crystal I have told. I would depend upon pellets for everything between the crystal and the last third of the filling, which should invariably be made with heavy foil.

My reason for this is that experience has taught me that fillings completed, or made throughout with heavy foil, will retain their smooth, polished surfaces much longer than any made otherwise. I have a theory to offer as a cause for this. If we take a strip of foil and roll it into a rope, there is necessarily air between the folds. If, now, we cut this into pellets, with each snip of the scissors we compress the edges of a pellet, practically sealing within it several cells of air. When filling with these, we only compress the gold as much as the contained air will permit. It will be claimed here that this is an argument, first, against using such pellets at all, and secondly, in favor of replacing them with the blocks or cylinders which I have condemned. My reply to the first would be that pellets of this kind, in spite of the argument, will make a good gold filling, even used at the surface, and where heavy gold is employed for the last third of the filling, they certainly serve all our purpose. To the second, while admitting that the block or cylinder has the edges open, thus appearing to offer an outlet for any contained air, they are nevertheless open to exactly the same objection, because to surely expel the air it would be necessary in condensing to begin at one point and then work exactly uniformly toward the edges of the pellet, or to start at one end and approach the other, neither of which is probably done in any instance, and surely not with every block or cylinder. The heavy foil is better because it is of a single thickness, and does not contain any air at the start. Yet, even with this magnificent material, care must be observed in the same direction, for if not properly attached and condensed, air-cells will be produced.

I now come to a discussion of the merits and usefulness of heavy or rolled gold. It is procurable in various thicknesses, numbered 30, 40, 60, and 120. The last is an extremity to which we need rarely if ever resort. Numbers 30 and 60 are sufficient for all purposes, and we may practically produce the latter by folding the former once, before cutting.

The earlier advocates of this form of gold were so enthusiastic over it that they claimed to depend upon it entirely. The theory was that the more gold one could crowd into a cavity, the better the result. This is not true. The supreme demand upon any filling is that it shall present a durable surface, and be in close contact with all its walls. It could be hollow in the center, and serve as well. Indeed, hollow fillings have been made, merely to show dexterity, and they have preserved the teeth. In large cavities surrounded with strong walls it is not very difficult to accomplish this. The floor and sides are covered with a solid but thin veneer, and then the top is roofed over by building from the margins toward the center, finally closing the last orifice, and leaving the interior of the filling hollow. It is not quantity of gold therefore that is needed. At the same time we must have solidity at the surface, as that means durability. This is best achieved when we depend upon heavy gold for at least the final third. A filling so completed will receive a higher and more beautiful finish, and will retain its luster longer than any other.

I have hinted that there are places, however, where we must depend upon this gold entirely. I will describe them in more detail. We have all of us seen, in crowded jaws, bicuspids so solidly rooted that it becomes almost impossible to obtain extensive separation. Approximal cavities in these teeth, especially where they reach almost to the gum, are most trying. The orifice at the masticating surface is so small that an ordinary pellet is compressed in passing, and yet, when within the cavity, is so loose that it is most difficult to attach it firmly against any gold already in position. Our manipulations only compress it more, when in the end most probably it becomes a nonplastic nodule unfitted for use. It is here that we may turn to heavy gold with entire confidence. Because of its thinness it can readily be passed between the teeth and led into place, when its impaction becomes comparatively easy. Undoubtedly a tooth thus filled entirely with this material will be well filled. If the question be asked, Why not do this in all cases? the reply is, that to do so would be a waste of time, as it takes longer to fill exclusively with heavy foil than where we use a plastic, pellets, or both, for the first two-thirds.

Another most important position for heavy foil is where in, let us say a lower molar, we find a cavity in the masticating surface, and another in the posterior approximal. Imagine this latter to be what we term a saucer-shaped cavity,—one of those exasperating places where we find absolutely no retentive shape, and a tooth so highly organized that every attempt to use the engine, or an excavator, causes excruciating agony to the patient. It is seen that if the two cavities be united, the whole would at once receive and retain a filling permanently. This is done, but now the question arises, At what point shall the filling be started? It is simply impossible to begin at the bottom of the approximal cavity, because there is no retentive pit or groove to aid us. We therefore begin at the wrong end, according to ordinary principles; that is, we begin at the top and work downward. We fill the cavity in the masticating surface, and then try to build over and down into the approximal portion. This is very difficult with any material, and trebly so with anything except heavy foil. The procedure is to only partly fill the upper cavity, and then, using No. 60 rolled gold, work over the edge, down into the approximal extension. The stiffness of the gold, and the fact that it is a single thickness, tend to make this comparatively simple. Here is a place where the foot-plugger is pre-eminently useful. With it a piece of gold may be first well attached to that already in place, and then the end led over the edge and against the wall of the approximal cavity. A second piece treated similarly will make sufficient stiffness to insure the overlapping part against riding up. Then a step further may be taken, and so on, until all the walls and floor of the approximal cavity are covered, when the completion of the work becomes simple with any kind of gold preferred. Care should be taken that as each new piece is added, it be first securely attached in the upper cavity. This is why I would not fill that part too full at the outset.

There is a most important use for heavy foil, in cases which are frequently very annoying. I allude to cavities at the labial festoon in anterior teeth. We not infrequently find cavities which defy the application of the best clamp at hand. Do what we will, there is a tendency on the part of the clamp to slip so that it passes over the edge and impinges within the cavity proper. In these cases proceed as follows: Get the clamp into place, and hold it there with one hand while the packing of the gold is begun. As rapidly as can be done with safety, fill along the upper edge until there is a good starting-point made, with either crystal gold or pellets. Then take No. 60 foil, cut into pieces about as long as the extent of the cavity and not too wide. Take one of these pieces and lay it upon the gold already fixed, so that about half of it protrudes above the gingival wall and rests against the clamp. Mallet this down only as far as it touches the filling, but leave the protruding portion. Be very careful, however, that the margin is properly covered. Repeat this three or four times. This will give us, say four thicknesses resting against the clamp. If this be now malleted down, a ridge will be formed, extending above the edge of the cavity, perfectly protecting the operator from any slipping of the clamp. The continuance of the filling will become a simple and pleasant task.

In very small cavities in the same neighborhood I have successfully filled with gold, without using the dam at all, by thus throwing up a barrier of gold, and then working rapidly. I have in some instances made this gold wall so high that at the end I have turned it down upon the filling and malleted it solid, though the upper side was wet from accumulated mucus.

The same general plan, though to a more limited extent, may be advantageously resorted to in those cases where, a ligature being an essential, it is found just possible to press it above the cavity margin. There is constant danger of its slipping. Use the heavy foil and form a ridge beyond the margin, as has been described, and there will be no further reason for anxiety as to the ligature.

Sometimes in a large cavity there is an underlying layer of oxyphosphate, either placed by the operator, or found in the cavity from which a leaking filling has been removed. For good reasons in the individual case, it may be desirable to fill with gold without disturbing this. Under the action of the mallet, occasionally a dust is formed by particles being broken from this layer of phosphate. This is not only annoying, but it interferes materially by preventing cohesion of the gold. The filling having been fairly well started, a single thickness of heavy foil may be attached, and then laid down, completely covering the phosphate, and the trouble is remedied.

There is one more use of heavy foil to which I will specially allude, and that will lead me to the proper method of manipulating it.

Once more let us imagine a cavity in the approximal surface of a bicuspid. While we have apparently sufficient space, or so judge at the outset, as the work progresses we wish that we had obtained more ; or let us say that more was unobtainable. The result not infrequently is that near the end we find that the cavity is perhaps filled so far as having its surfaces covered, but we wish to add more gold for the sake of contour ; the filling is flat, whereas we wish it rounding. The space will just admit a thin burnisher. In such a condition we find that there are parts of the filling, about its center, which cannot well be reached with a plugger. I proceed as follows, and I will say here that this is not to be confounded with the Herbst method. I lay a piece of heavy foil, preferably No. 30, within the space and against the surface of the filling. Then with a thin, flat, *clean*, and warm burnisher I burnish the added piece vigorously, whereupon it unites thoroughly with the filling. In a similar way I proceed, adding gold as long as needed, and completing my contour. This makes a hard surface, and though I have practiced this for over ten years, I have never had such portion of my filling scale off. Another, and perhaps more frequent, occasion for this method is where from long malleting, or other cause, a tooth is so sore that it is cruel to use the mallet any longer. I find that ordinary hand-pressure in this case results in no gain. Then I resort to the heavy foil and burnisher, with satisfaction to myself and patient. I would impress one fact upon the mind of the reader : *Never attempt this method until all margins are covered, and never try it with anything but heavy gold.*

I believe that the common practice is to cut heavy foil into narrow strips of considerable length, and fold it over and over as it is packed.

There is no objection to this where the operator has acquired the necessary skill. The beginner, however, will find it unsatisfactory, because it is difficult. In fact, heavy foil is the most troublesome of all golds with which to fill teeth. It must be remembered that it is the least pliable, and therefore we readily see that it will be found most unyielding in the hands of one who resorts to it for the first time. It must follow, then, that the less of it the operator has to contend with the more likely will he be to succeed. There is also a strong objection to using the gold in long strips. It will frequently be found that it does not turn over so as to assume the direction desired. Then when it is forced into proper position an angle is crimped up, which when malleted down forms a hill, and is very resistant besides. I therefore think it best to abandon the strip, as I would the rope, and rely upon pieces not much, if any, longer than the cavity. Of course there is not much objection to making one fold with it. Here, then, as with all the other forms of gold, I advocate small pieces. It may seem that this makes a man a slow operator, but in the first place it is better to be thorough than rapid ; secondly, thoroughness does not necessarily mean slothfulness ; and thirdly, it is a fact that these methods may be followed and yet result in rapid work. Even if it did not, the pleasure of seeing fillings five and ten years old, with beautiful smooth surfaces, will repay for the outlay of time, patience, and conscientiousness.

As with other golds, I prefer small points in packing the heavy foils, and certainly the foot-plugger is at its worst when we are using rolled gold. Except when using a burnisher as described, it is a difficult and slow process to pack it by hand. Better and more rapid work results with a mallet, and better still where the mallet is one with rapid stroke, as the electric or the engine mallet.

Before leaving this I will point out one disadvantage of heavy foil. Being of a single layer, it is difficult with it to alter the form of the surface against which it is packed. For example, suppose that in following and covering the walls of a cavity the center is left somewhat hollow. If now we begin to add heavy foil, the hollow will not be materially filled up. The best plan is to alternate with pellets, placing a pellet in the hollow, packing it, then using the heavy foil, then another pellet, and so on till the hollow is filled flush, when we may continue with the heavy foil alone, till the whole is completed. Another thing to be carefully guarded against is to allow a piece of heavy foil to become rumpled or crinkled when placing it between teeth. It will be found doubly resistant in such cases, and so stiff that a good result is hardly attainable. Use it smooth.

Gold in Combination with Oxyphosphate.—I come now to the description of a method of filling teeth which in proportion to its value has been too little considered. It is a general practice in many cases to interpose a stratum of oxyphosphate between tooth and gold. The ordinary custom is to allow the plastic to set, before proceeding with the insertion of gold. I touched slightly upon this subject when discussing oxyphosphate, but I have reserved until this time a fuller description of the method and the conditions in which it is indicated. There are many cases where dentine is excruciatingly sensitive while the tooth is being prepared for filling. I allude to teeth in which there is not a suspicion of pulp-exposure. I state dogmatically and emphatically, that *these hypersensitive teeth should never be filled with gold, allowing the metal to come in contact with the dentine.* To emphasize the point which I wish to make, let us briefly consider the cause of sensitiveness in dentine. Dentine is made up of succeeding stratifications of hollow tubes,—the dentinal tubuli. These tubes contain living matter; whether fluid, semi-fluid, or true nerve-tissue, is undetermined and for my purpose immaterial. The contents of these tubes lead directly to, and come into actual contact with the pulp itself. This pulp is highly organized, and traversed with nerve-filaments to such an extent that no part of it may be touched without responding by a painful form of sensation. Consequently, the contents of the tubuli rest against a tissue which reports painfully upon the slightest provocation. These in their turn, whatever their formation, have the power of transmitting impressions. Thus, when an excavator cuts across them, pressure is produced, and transmitted to the pulp, which responds at once. Now, the dentine is sensitive, or responsive in this way, in exact proportion to the relative size of the tubuli and their contents. That which has a large number of tubuli, with small diameters, will be less sensitive than that which has fewer tubuli of larger size, for here we find more living tissue, more power to transmit pressure, and consequently more pain. The patient says, "Doctor, the nerve is exposed, because you hurt me very much." The learned gentleman remarks, "No, madam ; there is no exposure. The dentine is sensitive, that is all." That is all! But microscopically speaking the patient is correct. *The pulp is exposed in hundreds of places. Every gaping tubule is an open passage to the pulp.* Hypersensitive teeth, those which I say should not receive gold directly against the dentine, are those where the tubuli are abnormally large. *To place against the open ends of these tubes a mass of metal of such conductivity as gold, is to invite, and often to induce, the death of the pulp.* These are the teeth where the patient returns and says, "Doctor, that tooth hurts whenever I drink cold or hot drinks." The dentist replies, "That will pass off." This is true, and it occurs in one of two ways. *Sometimes the pulp dies,* after which of course hot drinks do not cause sensation. Then, again, Nature may succeed in repairing the damage done by the operator. She, with her wonderful intelligence, at once begins a deposition of lime-salts against the surface of the gold. The tubuli thicken as to their walls, so that their contents and consequently their power to transmit impressions diminish. So much for the physiology of the condition. Let us now consider the treatment. It is to *smear the dentine with oxyphosphate, and while this is still plastic, and of sticky consistency, crowd in two or three well-annealed pellets of gold.* In large cavities crystal is better. The phosphate must now be left to set *thoroughly.* Then the gold may be compressed, and will be practically cemented into place. This aids in its retention, *but should not be depended upon to the exclusion of the usual methods of shaping the cavity.* The next step is to carefully scrape off all oxyphosphate which reaches or covers the margins. The filling may then be completed with any gold. It will be claimed that even oxyphosphate is a good conductor of heat. This is true, but it is less so than gold. Besides, the conductivity of a material depends upon its homogeneity. The fact that a filling of this character is composed of two masses, gold and oxyphosphate, renders the whole a poorer conductor than if it were made of either exclusively.

Beginning, then, from the point of using this method in sensitive teeth, we quickly determine to employ it in cases where the pulp is really nearly approached. Here a greater mass of oxyphosphate may be used. When we have become dexterous in the operation in these two classes of teeth, we soon begin to wonder why it is not applicable in those large or deep cavities of retentive shape, but where we find it difficult to obtain a starting-point. At once we see the advantage of using the phosphate *for cementing in the first pieces, merely to start a filling.* As it is perfectly feasible, is it not more sensible than to drill a pit, or bur out an extra groove? So that having passed through all the phases, beginning with ridiculing the method, I have almost reached the point where I use it exclusively. Where I do not employ it is mainly in cavities which are quite small, or where the immediate circumstances seem to make it less convenient or advisable than to depend upon gold alone. It is a habit that will grow upon a dentist, and the result upon the patients in his practice will be a lessening of the number of those reports about teeth being sensitive after filling. Of course there are some teeth which will be responsive to hot and cold, even after this has been done. Then the dentist may be comforted with the thought that it would have been worse if gold alone had been used. Indeed, the pulp would most probably have died.

Gold and Platinum.—This is a preparation furnished by some manufacturers. The exact proportions of the two metals I do not know, but there is enough of the platinum to materially affect the color of the finished filling, though during the operation there will be little, if any, difference in appearance. In the rough, the general surface will resemble any ordinary gold, but as soon as a bur or stone is passed over it the platinum asserts itself, and the yellow almost vanishes, giving place to a color which is not as beautiful as either gold or platinum alone. This matter of color I dwell upon, because to me it has been so objectionable that I have almost abandoned the material. I once had to build down the lower third of a living lateral incisor, broken off by a fall. For some reason, which I have never explained satisfactorily to myself, I used this combination. The patient was charmed, because her one dread was that she would show a lot of gold. I suppose it was mainly to humor this fancy that I did not depend upon gold alone. As soon as she stepped out of the chair and stood a few feet from me, I realized what a mistake I had made. The filling did not show as much as gold, because, except in bright light, *it did not show at all.* It looked so dark that practically the tooth seemed unfilled. To my mind, therefore, it is contraindicated for ordinary work in the front of the mouth.

An advantage which it possesses over gold alone is that it produces a much more resistant surface. That is, it is tougher, or, to express it otherwise, more dense. This quality would indicate its use where we find a whole set of otherwise sound teeth being worn away by abrasion during mastication. Some men chew so hard that they actually wear out their teeth. Ordinary gold fillings only very slightly check the havoc which is being done. If, however, we solidly fill all the molars with gold and platinum, further mischief will be prevented. For similar reasons it should be employed where, in cases of pyorrhea, it becomes expedient to unite two teeth by a filling extending across from one into the other, covering a platinum bar. The hardest possible filling is desirable in these cases, and is obtainable with this material.

Its manipulation deserves a few words of description by way of caution. In the first place, care must be taken not to tear the surface, as this exposes the platinum, and thus renders it non-cohesive at that point. For the same reason, annealing should be done in a way that will not risk burning off the gold. There are annealing apparatuses which serve very well, or a piece of mica may be used as a tray to carry the foil over the flame before it is cut up. When I am using this material, and I find that it is losing its cohesiveness, I lay on one or two pieces of heavy gold foil, thus producing the cohesive quality in the surface of the filling once more, and then continue with the platinum and gold.

Gold and Tin.—In spite of the high authority for this method, it is one in which I have no faith. In the first place, I think we make a great mistake in searching out as many materials as possible with which to fill teeth. We have now good and reliable filling-materials, which have served us well and will continue to do so, if we patiently and conscientiously apply ourselves to the acquirement of skill. I do occasionally use gold and platinum, and I have indicated where, when, and why I do so. Tin and gold does not appeal to me, because in the first place the want which it is supposed to fill does not in my opinion exist. It is said to last well where gold fails at the gingival margin. I have touched upon this subject before, and here will only state that I think failure at the gingival border is due to faulty manipulation, rather than to material. Certainly I should doubt that tin and gold would succeed where gold had failed. I made the statement once, before an advocate of this method, that I had never inserted a tin and gold filling, but that I had taken out quite a number of them, and they were not of that consistency described in written reports,—that, in fact, they had proven utter failures. The gentleman to whom I am alluding is a prominent and skillful operator, widely known, and has given many clinics upon this method. At this time he was filling a tooth while I looked on. In reply to my remark, he said that if I had taken out tin and gold fillings they must have come from the hands of poor operators, or at least men unskilled in this particular method. Of course to this I could say nothing. The patient for whom he was working was the office assistant of a dentist, a friend of mine, and I had thus a chance to learn the result in this very case. Within one year the tin and gold filling was leaking so badly that the tooth ached, so that it was removed and replaced with gold, which is still doing good service. I believe that a practitioner may be thoroughly successful, and satisfactorily serve all of his patients, without any knowledge whatever of tin and gold in combination.

Gold and Iridium.—This is a combination not manufactured, but procurable upon order. It is a piece of iridium placed between two layers of gold, and then the three rolled into one, producing what looks like heavy gold. It is very hard, and so difficult to manipulate that it barely deserves mention. The gold and platinum will serve every purpose for which this last combination was designed.

How to Condense Gold.—What we term "hand-pressure" is the first principle of packing and condensing gold into a cavity. It has several advantages over malleting, and some distinct disadvantages. Consequently it is rarely proper to depend upon it alone, and I might almost say the same of the exclusive use of the mallet. A dentist should be competent to employ all methods, and discriminating enough to know how to alternate them to advantage. The greatest good gained by hand-pressure is that the gold remains more cohesive under this method than in connection with any other. I am not prepared to discuss the physics of this phenomenon, but it is a fact long ago observed by me clinically, and I have sufficiently tested it to feel safe in making the following dogmatic statement : *The more gradual the pressure exerted upon gold foil in condensing it, the less it loses its quality of cohesiveness; and vice versa, the more sudden, sharp, or rapid the blow of the hammer, the less cohesiveness will be exhibited.* This is a very important statement, and being true, once recognized should prove invaluable to the operator. For example, suppose the Bonwill mechanical mallet, with a very rapid stroke, is being used in filling a cavity ; suppose that suddenly, seemingly without reason, the gold, piece after piece, refuses to cohere with that already packed. We stop and examine for moisture, but find none. Then we try another and another piece ; perhaps they fail, or they may cohere only to come away again after a few pieces have been added. What is to be done? Is it an uncommon experience? I think it will be recognized by many. If the reader, the next time he is in that predicament, will *pack two or three pellets carefully by hand-pressure*, he will possibly be astonished at the fact that they cohere perfectly, and that as soon as the whole surface has been covered with the new gold the mallet may be resumed satisfactorily. This is intelligent alternation of forces. Again, suppose a large cavity, with a comparatively small opening. It is found that before the floor is covered, the mallet seems to have a tendency to induce the filling to leave its retaining-points or grooves. The filling becomes loosened. Again and again this loose piece is removed and the work restarted, but always with the same result. Begin once more, and fill a good portion of the cavity by hand-pressure. After that, continue with the mallet, and all will be well. Suppose that we wish to use large pellets in a large cavity. Every now and again the pellet balls up under the mallet, or only one end of it coheres, so that we tear off and discard the other half. Here is a time when *each large pellet should be partly condensed by hand before the mallet is brought into use.*

The disadvantages of hand-pressure are obvious. It is slow, it is tedious to both patient and operator, and it does not produce as dense a filling. More important still, it is a fact that more teeth have been actually broken by hand-pressure than where judicious use of a mallet has been depended upon. To use a corkscrew plugger by hand in a deep undercut of an incisor, is to invite the fracture of the corner. The slightest twist of the instrument brings so mighty a force against frail structure that the disaster is inevitable. This is a lesson that students should learn theoretically, rather than practically, as I did. It is very unpleasant to try to palliate the offense to the patient. Again, a plugger is more likely to slip by this method than where the mallet is in use. The force exerted is a continuous one, and practically limited only by the resistance. Thus, if the resistance gives way, as when the instrument rests at an improper angle so that it may slide off the tooth, the pressure carries the instrument onward, and does damage. It may be only to tear the dam, or it may be to emerge through the cheek of the patient as I saw happen once, or it may be to pass completely through the guard-finger as occurred to myself. These little accidents, especially like the last, make an enduring impression.

The advantages of malleting gold are unanimously admitted, I believe. The only question remaining is, "What mallet shall we choose?" On this subject I have little to say. It is useless to discuss the merits of various mallets, for as many opinions can be obtained as there are mallets, or men using them. There are a few points in connection with the ones most used to which it may be profitable to allude. I was taught to use the automatic mallet. I was a strong adherent of it for a number of years. I feel satisfied that good work can be done with it,—as good as with any other. But I once had a tooth in my own head filled with it, and have never used it on a patient since. *There is one feature in the ordinary hand-mallet that is not to be found in any other: the patient is connected with the mallet only at the exact moment of contact as the blow is struck.* With all other mallets the patient is practically connected with the instruments behind the mallet all the time. This is especially true with the automatic. What is its action? The point is allowed to rest against the filling, and then pressure is exerted till a released spring causes a blow to be struck. Is it not evident that the whole blow is anticipated by the patient every time? Is it not plain that, to a nervous person, this expectation of the coming stroke must be maddening? It was to me, and I am by no means described by the curious word "nervous," which in fact should read "nerveless." With the electric and with the mechanical mallet it may be argued that the patient is made to feel the mallet only at the moments when the blows are produced, but these occur so frequently that the patient is practically connected with the instrument all the time. Besides, the force which manages the hand-mallet is that magnificent machine made by the Creator, which, without being oiled, runs noiselessly. The powers which control the two mallets mentioned, even when thoroughly oiled, make considerable noise, and produce an answering response from the suffering patient.

I think that the hand-mallet is preferable to all others. I depend upon it whenever I feel that there is necessity for the very best attainable result. Of course it has its limitations. There are places where a man would need three hands,—one for the mallet, one for the plugger, and one for a mouth-mirror with which to reflect light, or with which to see the cavity. For I must deprecate the habit of having an assistant do malleting. I cannot see how two brains can serve as well as one in this instance. Of course one must be ambidextrous to do this, but *all dentists should have the free use of both hands*. I do not hesitate to use my engine with my left hand as quickly as I would with my right. Why children are ever taught to acquire the habit of depending upon one hand, thus partly losing the use of the other, is one of the mischievous mysteries of civilization. It is one of the follies of fashion. A dentist who cannot use the left hand may acquire its use as I did, by practicing writing at all spare moments. Do not write as one does with the right hand, but reverse it; that is, begin at the right-hand side of the paper and write toward the left, or what we would call backward. This is because whenever we use a muscle a certain impulse to the same end is sent to the fellow of the opposite side, so that our long habit of writing with the right hand has, to a certain extent, educated the muscles of the other hand to the same purpose. When we use the right hand we begin on the left side and move toward the right; therefore, to get the same action from the muscles when we use the left, we must work from right to left. In a short time one can learn to mallet with the right hand, and manage the plugger with the left. This is more satisfactory than depending on the assistant, who may or may not mallet to suit.

Of the power mallets, I prefer the Bonwill mechanical, as being easier to manipulate and more satisfactory in its working. Certainly there are no batteries or motors to keep in order, or add to the noise. The stroke is always under control, either as to speed or force of blow. There is one thing worth mentioning in this connection. When I first obtained a Bonwill mallet, I did so only on trial. I think I was attracted to it because of its ingenuity rather than by any inherent merit which I expected to find in it. I was very much prejudiced against such an instrument. All the argument which I have just used about the connection between the patient and the mallet, I had powerfully before my mind. I tried it conscientiously, however, with the result that by alternating between it and the hand-mallet I obtained an expression of preference from my patients. I think fully ninety per cent. chose the Bonwill. The others were of course those nervous, or nerveless, individuals to whom noise becomes an irritant.

Pluggers to be used.—This is a subject upon which a great deal of discussion has been devoted, and I think largely wasted. The size and shape of a plugger must depend largely upon a man's personal preferences. Instruments with which one man operates beautifully would be worthless to another. Of course, certain mechanical principles can be cited, which theoretically will prove that there is but one form of point that best condenses gold, but in practice a broken instrument is as good as any, in the hands of the skillful. Points that are serrated, or round, or flat, or smooth, or thick, or thin, may either or all be made to produce good results. As to shape, a few varieties are convenient, but I may mention that when I first took the Bonwill I received only one point with it. I used that one point for a year, and though I have bought a few others since, so as not to appear stingy, I rely upon that same point now almost to the exclusion of all others. It is practically universal as a mallet-plugger. The young dentist should endeavor to get along with as few instruments as possible, rather than with many. With a few instruments the workman is the master; with many he is the slave, for he is powerless as soon as one is mislaid.

CHAPTER IV.

The Relative Values of Contour, and Flat or Flush Fillings—The V-Shaped Space in its Relation to the Gingiva—The Restoration of Superior Lateral Incisors—Slight Contours—Regulation of Teeth by Contour Fillings—Departure from Original Form—True Contouring—Treatment of Masticating Surfaces—Contouring with Gold—With Amalgam—With the Plastics in connection with Gold Plate—Use of Screws—Cases from Practice requiring Odd Methods.

To imitate nature should be the aim of every true artist. Yet the landscape-painter, who contents himself with copying what he sees spread out before his view, would frequently produce a picture which lacked composition. In dentistry the skilled workman is he who has the eye to see contour, the trained fingers to reproduce it, and the *judgment to decide when and where it should not be attempted.*

The reproduction of the original form of a tooth, which has been partly destroyed by caries, must be undertaken in over ninety per cent. of the cases presenting. This being my opinion, and being one contrary to the teaching and the practice of some men of eminence, I shall state some of the reasons for my conviction.

It will scarcely be disputed that the whole scheme of the universe is a perfect plan. No improvement upon it is possible. The construction of every creature is accurately adapted to his intended mode of life. Yet it has been claimed by some that the formation and structure of the teeth of man are less perfect than those of the lower animals, since man is the only animal who constantly suffers from caries. This disease has been found in other animals, but it cannot be said of any other species that we can take any individual at random, and be almost sure to discover the ravages of caries, as we do in man. Consequently we must admit that the human race is more susceptible to it. That this is due to the formation of the teeth themselves, or to their arrangement in the jaws, however, cannot be claimed, because the chimpanzee, the gorilla, and other members of the ape family have jaws practically similar to that of man, and yet caries is so rare with them that a jaw with the teeth showing it is a curiosity to be preserved and placed in the glass case of a museum.

The probability is that man suffers because he cooks his food, for we find that where we take animals unaccustomed to such diet, and domesticate them, feeding them upon our style of food, they frequently suffer from caries. The question then arises, Shall we alter the shape of a tooth which has been attacked? Shall we depart from the standard set up by Nature, because of the fact that we do not eat the kind of food for which we were designed? If we examine the dog, we discover that with his pointed teeth, although he eats from our table, he suffers very rarely from caries. Shall we decide from this that we may file the human tooth into an approximation of a cone, producing such spacing that the tip of the tongue may remove the refuse food from between the teeth? This is a serious question, especially as the production of this V-shaped spacing was once strongly advocated, and largely practiced. It is indeed the custom of some dentists to-day, and it may come up again within a few years, just as many other discarded practices have been rediscovered, and taught and adopted, till their mischievous results once more condemned them to oblivion.

The V-shaped space will succeed in the human mouth, only when the gum of the human individual approximates in density that which we find in the mouth of the dog. In man the teeth occlude squarely, the one against the other. We grind our food between enamel-surfaces. With the dog, except in the incisive region, the teeth pass between one another, and *bite against the gum itself.* This gum-tissue is comparatively thin, and supported below by a dense alveolus. Such a thing as an inflammation, or hypertrophy, in this region is unknown. I have seen pyorrhea alveolaris in the mouth of a dog, but that was in the incisive region. How is it with man? The gum-tissue is arranged in a very different manner. We find between all the teeth a pedicle extending toward the incisive edges. *This pedicle is the thickest part of the soft tissue, and therefore least able to withstand irritation from pressure.* Any attempt to chew upon it will result in disease. In the normal mouth it is protected by the approximation of the adjacent teeth, which by touching each other are expected to prevent food from passing between the teeth, and impinging upon the gingiva. I argue therefore that we cannot allow ourselves to copy the dog, because while we may give man the dog-shaped tooth, *we cannot assure him a resistant gum-tissue between the teeth, which will withstand the crowding of food against it,* as occurs with the lower animal.

In conditions of disease where pyorrhea alveolaris has been present, we have been taught, and well taught, that until we can produce rigidity of the teeth, we cannot hope to see the disease cured or even temporarily controlled. Why is this? Because as long as the teeth are readily movable, the soft tissue will be in a constant condition of irritation due to such motion. If this be, and it is true, should we not pause before producing or leaving spaces between teeth? As soon as a space is made, do we not destroy the integrity of the abutments in the arch, and so make mobility of all the teeth more possible than before? This is especially true of the two teeth on either side of the space. Whenever food is chewed in that region, it is packed between the slanting planes, and acts as a wedge to drive the teeth apart. Not infrequently this results in a loosening of the two teeth, accompanied coincidently by an inflammation of the gingiva. This inflammation goes on till a suppurative stage presents; then there may occur hypertrophy, and often pyorrhea directly results. I have noted mouths where several teeth were affected with this dreadful disease, and then have found that the teeth most involved were two between which the V-shaped space had been made. I have no hesitation in saying that in these cases the pyorrhea was first caused as I have specified, and that subsequently the neighboring teeth suffered by infection. Thus the dentist who filled the space invited, and I may say induced, pyorrhea. Of course these are extreme cases, and there are instances where the V space has been entirely successful; but in all cases that have come under my observation, and they have been very few, I have noted that the gum-tissue was dense and tightly drawn over the alveolus, the pedicles between the teeth being of a cartilaginous firmness, short, and tough.

In brief, my chief reason for advocating contour fillings, aside from any consideration of cosmetic effects, is that to leave a space, or to form one, between two teeth, is liable to result in a loosening of the teeth, or a diseased condition of the gingiva. In the latter case, caries will almost certainly reappear along the gingival margin of the cavity, and undermine the filling.

The production, then, of a perfect contour, is the most important element in the successful filling of teeth. A novice may soon be taught to stop a small cavity which is surrounded by strong walls. To insert a contour filling, of such a form that beauty, usefulness, and strength shall each be attained in the highest relation to the conditions presenting, requires a skillful manipulation, a knowledge of tooth-form, and a mature judgment, which only experience can give. I can only hope to direct attention in the proper channel. No theorizing can give the student such attainment.

In every case which presents, the question will arise, Can the original form be reproduced, without danger of future failure? This requires often a keen knowledge of the mechanical arrangement of the cavity for the retention of the filling, of the probable strength of the special tooth under consideration, of the force of leverage which under mastication the filling will exert upon its anchorage, and of the amount of usage which the particular patient will give to it. There are men who chew with such energy that they not only wear out their teeth early by abrasion, but the best gold filling will become roughened rapidly in their mouths. In such cases it would be folly to attempt an extensive contour with doubtful anchorage; whereas in the jaw of a refined and delicate woman, the experiment might be made with much safety.

The lateral incisor of the superior jaw offers the greatest problem in contour, but fortunately it is a tooth so variable in shape that we may be enabled in almost any case to so fashion the filling that it will be durable and at the same time sightly; that is, if compelled, we may produce a contour which is of good form, though not similar to the original structure.

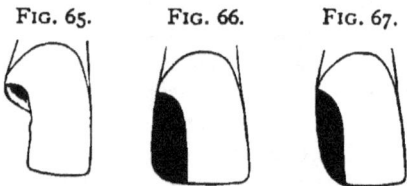

FIG. 65.   FIG. 66.   FIG. 67.

Fig. 65 shows a lateral incisor with a cavity which is often puzzling. Where the pulp has been removed, we may obtain such good anchorage in the upper part of the cavity, packing the gold well into the canal if need be, that we would have no hesitation in producing a full contour as seen in Fig. 66. But suppose that the pulp be alive, and despite the ravages of caries sufficiently well covered by dentine to render its salvation assured? It would be most reprehensible to devitalize it. Yet to attempt to obtain an anchorage which would retain such a contour as shown in Fig. 66 would in most cases be impossible, and even where attained, we would be likely to approach the pulp so closely at some point that it would be liable to die subsequently. I will give a case of this character from practice, which will be instructive. But first let me dispose of this lateral incisor with its living pulp. We may fill the tooth, preserve the life of the pulp, and produce a presentable and durable contour, by resorting to a screw, and shaping as shown in Fig. 67. The distal corner of the lateral incisor is often a fairly sharp angle, but it is more often a curve, and sometimes as much curved as shown in the last figure. If Figs. 66 and 67 be compared, it will be seen that in the former the length of the gold along the incisive edge

RESTORATION OF INCISORS.

produces a strong leverage, which during mastication would operate to force the filling out of the cavity. Where the anchorage therefore is slight, it will be safer to adopt the second form, or as much of an approximation to it as judgment shall direct, for the leverage is almost annihilated by the curved line, while we still retain a semblance of tooth-form. Where the tooth of the opposite side is so markedly different as to make too decided a contrast, there might be no harm in slightly altering it so as to produce harmony.

To come to the case from practice alluded to. The patient was a married woman about twenty-five years of age, and decidedly anemic. She was excessively nervous, and her teeth hypersensitive. The history given to me was that the two teeth which she wished me to fill had already been filled seven times, each time with gold, and all had failed. They were a central and a lateral incisor. The first examination showed two crumbling gold fillings, of very poor shape and leaking badly. They were readily tipped out with an excavator, and the cavities were most uninviting. The central incisor seemed to be the more difficult, and therefore was undertaken first. Fig. 68

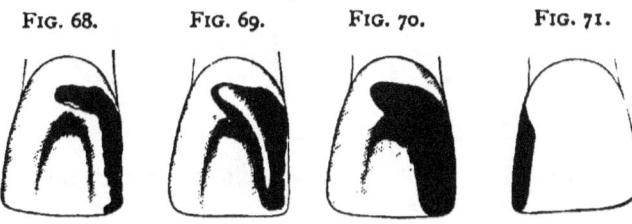

FIG. 68.   FIG. 69.   FIG. 70.   FIG. 71.

gives the condition of the tooth seen from the palatal aspect, where the loss of tissue was most apparent. After crowding the gum away by packing cotton against it for several days, so that the dam could be forced above the gingival border, I prepared the cavity as seen in Fig. 69, making a slight groove immediately within the border at all points, and extending one or two horns of the cavity upward, avoiding the pulp. As will be observed, this practically surrounded the pulp with gold between which and it there was but little dentine. To make all more sure, I used the method previously described, covering the cavity with oxyphosphate, before inserting the first pieces of gold. I then filled the tooth so that it appeared as shown in Figs. 70 and 71, which give the palatal and labial aspects. Observe that very little gold shows from the labial point of view, and that though the anchorage here was very poor, it was probably sufficient, because the leverage is very slight, there being but little gold along the incisive edge. I felt quite elated at my success in this most difficult case, and turned to the lateral incisor with considerable assurance. Here I found so much decay that I was compelled to destroy the pulp, which was exposed badly. This, however, gave me a better anchorage, and I anticipated no trouble. Immediately after the pulp was removed, the patient reported continuous and excessive pain at night. I did everything to allay the mysterious suffering, and it was not till a week later that she was able to assure me positively that it was the *central* incisor which ached, and not the *lateral* incisor as we both had supposed. I then reluctantly concluded that the presence of so much gold so near the pulp, even protected by the phosphate, had resulted in a pulpitis. I drilled into the palatal side, entering the pulp-chamber, and after considerable hemorrhage removed the pulp without resorting to arsenical treatment. I am as positive as one can be that this pulp was not exposed even minutely, as we understand the term exposure. That it was very nearly approached there was no doubt, and it is plain now that it would have been better to destroy it, thus obtaining a stronger anchorage for the filling and rendering the work far less

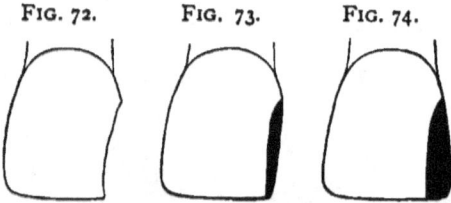

FIG. 72.   FIG. 73.   FIG. 74.

difficult. It was a case where a screw would have rendered the operation excessively tedious, because of the position of the cavity, which was almost out of sight. It was a cavity the difficulties of which cannot be appreciated from a description, even accompanied by the best illustrations, and I introduce it only to show that the immediate approach of gold to a pulp is a menace to the life of that organ. The phosphate did not prevent the disaster, because it was removed from all the grooves and deepest parts of the cavity, that the filling might be more secure, and it was in these places that the gold approached the pulp nearest. After the removal of the pulp all pain disappeared, proving that this was the offender.

Where the approximal surface of an incisor has been only slightly lost, many men, possibly because they lack a knowledge of tooth-form, make no effort to restore contour. The eye of the true artist finds no difficulty in designating what was the original form of a given decayed tooth, for he judges from what is left of the lines, and by continuing these in his imagination has in his mind a picture of what existed before the destruction. In Fig. 72 we observe a central incisor the approximal portion of which has been lost by caries. I have seen many cases of this character filled so as to appear as seen in Fig.

73, and even filled flat, so that no gold at all was visible from the labial aspect. There may be instances where such a course would be permissible, and I am not altogether discouraging the practice of not showing much gold. The point that I would bring out is that Fig. 73 does not restore the original form of Fig. 72. The eye of the artist must see at a glance that the lines of the filling in Fig. 73 indicate a different form to that shown in Fig. 74. The production of these very slight contour fillings always indicates the hand of a master. When I see a delicate line of gold showing, which just supplies the lost tissue, I inquire the name of the dentist and place a mark to his credit. I do more than that, for where he is resident of a different city, I place him on my list of those to whom I refer patients of mine who may be traveling.

The restoration of these slight losses along the approximal surfaces brings me to the consideration of a subject which I have seldom seen discussed. In crowded jaws we sometimes see two teeth, let us say the superior central incisors, each decayed along the mesial surface,

FIG. 75.   FIG. 76.

and yet in close contact. A glance seems to indicate that they have actually been pressed into each other. Such a condition is shown in Fig. 75. What has occurred here? As teeth have erupted posterior to this region, the decay along these surfaces has allowed the central incisors to yield and be crowded together. If they are filled with flat fillings, as in Fig. 72, we do not show gold, but we leave the teeth in their irregular and unsightly position. These teeth can be regulated, and retained in their proper, erect position by simply wedging them far enough apart to permit a restoration of contour with gold. They can be made to appear as in Fig. 76, which is assuredly more pleasing.

There is a departure from normal contour which is advisable, especially in the bicuspid region. In Fig. 77 we noted the normal condition, and find that though the point of contact is not very great, there is no space for the accumulation of food and *débris* because the gingiva fills what would be a space. After caries sets in, this pedicle of gum is often lost, and the teeth come to us as shown in Fig. 78. If we restore the original contour precisely, however artistic that might be, it would not be fortuitous, for it would often invite a recurrence of caries along the gingival border of the cavity. In filling such teeth it becomes necessary to produce a form which, while sightly, *will allow contact along as much surface as possible.* Fig. 79 shows teeth so filled, and a comparison with Fig. 77, while showing that we have materially departed from the original structure, indicates that we have nevertheless retained the relation between the point of contact and the gingiva, whereas with a perfect reproduction of contour we should have left a dangerous space, in which food would accumulate despite the knuckling. Here again is another contention against the use of the matrix. Where that instrument is used, especially where the teeth

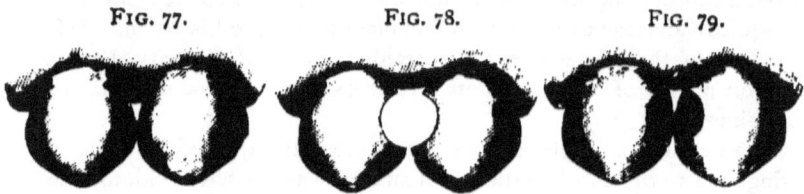

FIG. 77.   FIG. 78.   FIG. 79.

are not first separated, how can we possibly control the form of the filling so as to assure contact along the greatest possible surface? To me it seems simply impossible. If the matrix is not useful in these, the most difficult cases, why should we need it in those which are less perplexing?

To the minds of many dentists, the word contour seems to convey an idea only of the restoration of shape so far as concerns that part of the tooth which is exposed to view. I have seen approximal fillings beautiful in form and finish, *if looked at from the labial view*, but which at the palatal, incisive, or approximal aspects by no means restored even an approach to contour. Fig. 80 shows such a piece of work, all the imperfections of which are easily seen if we examine from the palatal side. We note that the palatal concaved surface has not been reproduced, that the approximal surface is not rounding and full, and that the incisive edge is not square, but rather sharpened to a thin edge. When I see this sort of work, I judge that the dentist is aiming chiefly to please the eye of his patient, rather than to produce a perfect contour, if indeed he has the skill to fashion what he evidently has not conceived. *To restore by contouring means to restore form from all points of view.* It is as essential at one part as at any other, for the cosmetic effects are of the least importance, though to be highly considered. In other words, we do not restore contour for beauty, but for utility.

FIG. 80.

Thus we come to the grinding-surfaces of bicuspids and molars. Shall we feel bound to reproduce the sulci when we form almost the entire surface with our filling-material? In bicuspids it is usually both advisable and necessary to do this, at least to a considerable extent.

There is no object in burring out two deep pits such as are found in the normal crowns, but as the occlusion in this region is usually very accurate, the cusps of the lower opposing teeth biting squarely and sharply into the sulci of their antagonists, it becomes essential in forming the filling to remember this, and to accommodate the form to the requirements. In molars I think that a middle ground is safest. We should not make an extensive gold filling and leave it absolutely flat, for we thus offer a very poor masticating surface. At the same time, to attempt to carve out an exact reproduction of sulci as seen in a real tooth would not only be time unnecessarily spent, but the result would be really bad. I am certain that gold thus carved will not be as durable as where a mere approximation is aimed at. Such depressions as may be produced with *fairly large-sized gold-cutting burs* will leave a surface sufficiently cusped for service without making it difficult to produce a perfect polish, which is essential to durability.

This brings us to a consideration of cases where, by the attrition of energetic mastication, a patient presents with a complete set of teeth, all of them having lost about one-third of their original length. Is it necessary here to attempt a complete contour? I think not. Such a person has gradually become accustomed to having the jaws close together more than when the teeth were perfect, therefore there is no discomfort to him because of the loss in length. The necessity for interfering at all arises from the desire to prevent further destruction. In tipping a whole set of teeth with gold in this class of cases, it is only necessary to build on a sufficient thickness of gold, so that what is placed may be strong enough to prove resistant. If a molar on each side be treated in this way at the first sitting, the mouth will be opened a trifle all round, and the other teeth in turn may be built up to suit the new occlusion. In the incisors it will be found that the cutting-edges are quite broad, so that there is little difference between them and the cuspids and bicuspids. No attempt should be made to produce cutting-edges similar to what *was* in the first place. What *is*, is what the man is accustomed to, and he will be happier if his teeth are left to him as they are. Of course the filling following the lines of contour may slightly narrow the edges, but this should not be carried to an extreme.

Many cases will occur in every practice where any attempt at contour will be impossible because of the fact that the position of the tooth has changed since the destructive process began. Often we see teeth lifted from their sockets, as the decay advances, so that an occlusion is effected. To build up such a tooth to original proportions would be to open the mouth, by making that the only tooth which strikes. In these and other similar conditions judgment must have sway, and direct the hand. There will be many cases, too, which will allow only a flat filling, and in some a flat filling, while serving every purpose, will entirely alter the original form of the tooth. Fig. 81 shows an approximal view of a bicuspid, the palatal cusp and side of which have been completely lost. In such a case, if the pulp is dead, a complete contour should be attempted, anchorage in the root-canal making this perfectly safe. The result is shown in Fig. 82, the extreme edge of the remaining natural cusp being ground off to

FIG. 81. FIG. 82. FIG. 83.

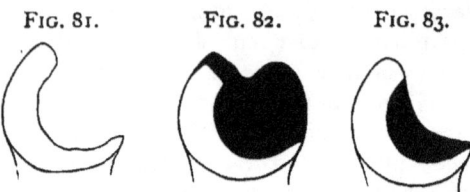

allow a masticating surface of gold only. Where the pulp is alive a totally different course should be followed, as shown in Fig. 83. Here we see that a flat filling has been resorted to, the tooth assuming the form of a *cuspid* rather than of a *bicuspid*. In this case the tip of the natural cusp is not removed, for with the shape in which the tooth is to be left mastication will not be a great menace to it, as it will where a mass of gold is inserted so as to reproduce the palatal cusp, which acts as a powerful lever to shatter the standing wall.

METHODS OF PRODUCING CONTOUR FILLINGS.

Gold.—When attendant circumstances do not contraindicate its use, gold undoubtedly will give the greatest satisfaction where any considerable portion of a tooth must be reproduced. I have known men who have claimed that they could make a permanent contour with *non*-cohesive gold. I do not doubt that such men do what they claim, but these individuals are very rare. Therefore I advise the beginner to depend upon *cohesive* gold when his filling must extend beyond cavity-borders. Moreover, I would suggest that he obtain gold *as cohesive as it can be made.*

There are certain methods of manipulating gold, essential when contour is to be reproduced, which may not be so in ordinary cavities. For example, I have said that a filling might be made hollow, provided it touched the walls at all points. This would be a grand error in a contour filling. It has been stated by some that a large cavity may be well filled with crystal gold, the lower portion being only partly condensed, provided the upper third be made solid. However this may be in ordinary work, it is not to be thought of in contouring. The rule must be that in placing a contour filling, every piece of gold, from the very first to the very last pellet, must be thoroughly condensed, and each and every pellet should display perfect cohesion.

METHODS OF PRODUCING CONTOUR FILLINGS. 93

To take up these two points separately for a moment : Let us suppose an ordinary cavity with surrounding walls ; it is half filled ; the operator places a pellet, and mallets it less thoroughly than he has done its predecessors. He adds another, and continues the malleting. Of course the force of the blows will further condense the under pellet. Even if it be not completely condensed, the fact that the shape of the remainder of the cavity is retentive without regard to that part already filled, makes it a matter of no moment whether that one pellet is, or is not, thoroughly packed. With a contour filling it would be a most hazardous oversight to leave even a single pellet of gold insufficiently condensed. That point would be a weak spot. However solid the rest of the filling may be made, the time may come when, under strong pressure in mastication, a fracture of the filling will occur just where that pellet was placed. If heavy foil is being used, the rule is even more imperative. Every layer must be thoroughly condensed before the next is added, for the reason that, with this kind of gold more than any other, an underlying layer is less likely to be condensed where another is superimposed before the malleting is completed. For a similar reason, no matter how much need there may be to hurry, the dentist should never pick up two or three pieces of heavy foil at one time and attempt to condense them in that form ; the ends, being irregularly arranged, will fold one over the other in such shapes as to offer the greatest possible resistance to the mallet, the result being improper condensation.

With the other point, relative to cohesion, the necessities for extreme caution, extending to every pellet, become evident along the same lines of argument. It is as bad for a filling to fracture because one layer did not cohere, as because there was a flaw from lack of solidity. As before, non-cohesion, or slight cohesion, may be overlooked in the body of a cavity having surrounding walls, because what is placed above it will still be retained by the upper part of the cavity. It is otherwise with the contour filling. If only one layer, especially if it be of heavy foil, fails to cohere, all that which follows is but added to, and is not a part of, the filling. A fracture may be expected at any time.

It therefore follows that the size of the pellets, or strips of heavy foil, should not be increased near the end of the filling in order to hurry the work. Larger pellets or pieces will render solidity and cohesion both less liable. Above all things, large or even moderate-sized foot-pluggers are to be avoided, though more permissible with the Bonwill mechanical or the electric mallet than where a hand-mallet or hand-pressure is relied upon. I wish to condemn the foot-plugger for this class of work, yet must speak with caution. Many men of skill use the foot-plugger with success, and with more rapidity than where another form was chosen. But these men select a foot-plugger which is narrow, and reaches a sharp point at one end. Thus in one instrument is had the action of a foot-plugger, or of a point. Such a plugger is shown in Fig. 84, and is most useful. What I am advising against is a broad, flat, and unusually long foot-plugger. This condenses so much surface at once that thorough cohesion is doubtfully, if ever, obtained.

FIG. 84.

One more essential point : In packing any filling I make it my rule that, from the very initial pieces, the shape of the cavity must be such that I can use the mallet, without needing another tool to hold the gold steady. This is the rule : exceptions exist, but are very rare. *Applied to contour fillings, the rule must have absolutely no exceptions.* Every piece of gold as it is added, must produce a filling, as far as the work has progressed, to dislodge which would require the engine drill. How this may be attained in specified cases, and in cavities offering special difficulty, will be discussed later. I merely call attention here to the fact that, in contouring, this rule must hold complete sway. If at any time it is found that the filling will tip, or move in the slightest, the operator may as well remove his gold and reshape his cavity. To emphasize this point, I will relate a case which occurred early in my career.

FIG. 85.

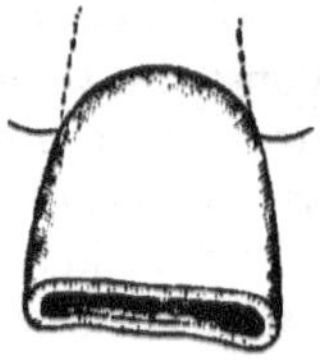

I was asked by a dentist, whose experience and skill were much respected by myself, to be at his office one afternoon to assist him in placing a large contour filling. He wished me to mallet and pass the gold. I did so. The tooth was a central incisor, and the cavity as he prepared it is shown in Fig. 85. It is seen that the cutting-edge is absent. The operator had prepared a cavity retentive in shape, merely because with a wheel-bur he had made an undercut, or groove, all around. This is exactly the plan to be followed in similar cases, where from abrasion the ends of teeth have been worn away, and it is desired to stop the destruction by offering masticating surfaces of gold. Then the filling is made *flush with the top of the tooth*, and will remain in place even though it must be held with another instrument all through the process of packing. (This, of course, need not be if the groove be properly shaped to retain the first pellet.) To so arrange a cavity, however, where, as in the case which I am citing, *about one-quarter of the length of the tooth was to be restored*, was absurd. Thus I thought, but of course made no comment. The dentist began with a rather large pellet, and proceeded much as though he had been using *non*-cohesive gold. That is, he was mainly depending upon wedging the pieces across from one groove to the other, using one instrument for condensing under the force of the mallet wielded by myself, and another to prevent the gold from rocking or moving. This was kept up till the whole cavity was filled flush, when of course it *appeared* solid. Then the work progressed rapidly, till the whole contour was completed, at the end of two hours' work. *In finishing this filling with a file, the dentist succeeded in straining it from its anchorage so that it could be slightly rocked.* He looked at me silently, and I refrained from speaking. He tipped it out, and, strange to say, proceeded to refill the cavity without change of plan. This time he succeeded in making the contour, and also in polishing it so that it *looked* really handsome. Moreover, it lasted as long as the lady lived, though I should record that *she died two weeks later.*

The essential features of a gold contour, therefore, are extreme solidity, extreme cohesion, and extreme immobility throughout and at every stage of the operation.

Amalgam.—It is frequently admitted that amalgam has been a much-abused material. This charge against dentists is more true in relation to contour fillings than in any other connection. It will not suffice to say that, for contouring, amalgam is a *valuable* agent. It is necessary to say that it is *invaluable.* Its usefulness is inestimable. It may be made to save teeth which without it would be lost, or, at least, even if saved, would be of but slight service for mastication.

In the realm of contour work, amalgam occupies a place that is unique. With it can be restored all those forlorn cases, those wrecks, which half a century ago were inevitably consigned to the forceps. Yet, with shame it must be admitted that only a very few men know how to obtain the most desirable results with amalgam in these very cases. The man who can restore a molar where caries has advanced beneath the gum, two or three cusps being entirely absent, and build upon this unpromising foundation a tooth which becomes as useful as the original, and which, moreover, remains permanently in place without fretting the gum and without inviting decay along its borders, has more right to count himself skilled than the best gold-filler that we have known.

In small cavities the plastic is the more manageable material, but as the size of a cavity increases, manipulation with gold becomes *less* difficult, the added obstacle being only the tediousness of a lengthy sitting. With amalgam it is otherwise, for the larger the cavity the *more* difficult it becomes to attain the highest success.

Amalgam, then, in contour work may well attract our special attention. I must point out the obstacles to its proper use, and tell how to combat them. How often have we all expended a half-hour or more restoring a lost corner of a molar, only to have it return on the following day, with a portion of it missing? We say to the patient, "You must have bitten something on that before it was thoroughly hardened," but that is no satisfaction either to patient or dentist. The work remains to be done over, and discussion does not mend matters, especially as the same risk must be taken again.

This tendency to fracture, in an amalgam filling, is due to several things. Of course, if the occlusion be sharp, the explanation given by the dentist may be true; mastication may have dislodged a portion of the mass. But where such an accident is possible, the dentist must note the fact and guard against it in advance. The filling must leave him so shaped that it will not be disturbed by the closing of the jaws, even so much as by the production of a slight scratch. Moreover, this immunity must be determined, not alone by the perpendicular action of the jaw, but by the lateral as well. The patient must be asked to gently move the jaw from side to side, as he would do in eating. This brings the cusps of the opposing tooth, or teeth, into all the different relations which they are to bear; and if the filling is unmarred by this, the single warning to chew upon the opposite side during the succeeding day, if obeyed, will bring the filling back in good condition.

But it is often by a cause other than mastication that the filling is broken. We take the utmost care to keep a gold filling free from moisture, yet some men do not hesitate to insert amalgam with the cavity and surrounding parts flooded with saliva. This so-called submarine work not only should not be practiced, but a clinician showing it before assembled students, or practitioners, should be roundly condemned. In contour work with amalgam it is of the utmost importance that the perfect crystallization of the mass should not be interfered with by moisture. The filling should be kept dry throughout the whole operation, if possible. Where the cavity extends far beneath the gum-margin, the tooth may be filled in two operations, though at the same sitting. Using the napkin as a dam, amalgam must be packed until it extends beyond the margin, sufficiently far so that the rubber-dam may be placed. Then the cavity, and the amalgam already in place, may be dried and the filling continued. Thus it is shown that because a cavity cannot be kept dry with rubber-dam from the outset, that is no reason why the filling should be allowed to become inundated several times, through a vain effort to control moisture with a napkin.

The next important point is to avoid fracturing the mass during the operation. This involves simply the proper application of force, and the proper consistency of the material. Amalgam, for use in a large contour, must be prepared slightly more plastic than for ordinary work. It is to be packed with balls of bibulous paper, with a wiping motion, thus forcing the material against the cavity-walls, as long as this action can be carried out. By this course the excess of mercury is forced out, and crystallization begins at once. Thenceforth the particles of that portion of the mass already in position *must not be disturbed.* If by exerting force in a wrong direction a part of the filling is fractured off, *it is folly to hope to get a good result by patting it back with a burnisher. The reunion will not be as strong as was the original union.* It would be better, where the accident does occur, to remove the separated piece, replacing it with freshly-mixed material. To give accurate direction on this point I must resort to a diagram. Fig. 86 shows a large contour partly completed. The dotted line shows the extent to which it is to be carried.

Whether the remainder of the filling be packed with bibulous paper or with burnisher, it is from this point on that fracture is to be feared. The rule is very simple. *Pressure must be exerted only in line with the greatest resistance, offered by the tooth itself.* To pack the amalgam in the direction indicated by the arrow *a* would be safe, whereas to follow the direction shown by the arrow *b* would be to invite failure.

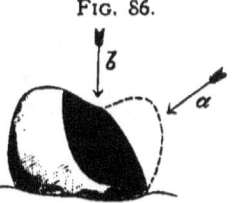

FIG. 86.

The last essential is to dismiss the filling as far advanced toward crystallization as possible. This may be best accomplished by burnishing gold into it, as has been previously described on page 57.

Oxyphosphate of Zinc.—This is a material rarely depended upon for extensive contour work. Yet there are two special conditions in which it may be made extremely useful. I once was applied to by an actress who had lost the corner of a central incisor. The tooth was unusually large, and the missing portion extensive. A gold corner would have been very noticeable to her audiences, and would not be tolerated by herself. It is possible that at some time in the future porcelain fillings will have been brought to such a stage of perfection that a case of this kind can be readily handled. The lady, however, came to me long before such work was even advocated. She simply asked me to restore the tooth with what she termed "bone-filling," and I did so, using considerable care to obtain as perfect match to color as possible, and succeeding fairly well. I replaced this in less than a year. It must be remembered that as soon as a small part of the mass had disintegrated the contour was obliterated, so that refilling in this position would be needed more frequently than ordinarily. The third time that I was asked to fill this tooth, I observed that the wasting away was mainly from the palatal side, outward. This compelled me to think a little, and I devised a mode of procedure which I have since followed in similar conditions, with most gratifying results. I first restored the shape of the tooth as before, using an oxyphosphate. This done, I burnished a thin piece of tin over the palatal portion of the filling, extending it partly around the approximal surface, and over the cutting-edge, trimming it to shape. With this as a pattern, I cut out a similar piece from thin gold (24-k.) plate, and treated it in the same way, thus fashioning a tray which would hold the material, and protect it wherever it was covered. This bit of gold was then soldered where the two turned edges came together at the angle of the corner, and a thin layer of solder flowed along the inner side. Into it were then dropped stray bits of gold, or platinum, and when heated up once more these were caught and held by the solder, producing a roughened inner surface. The filling was then removed entirely from the tooth. Its first insertion was intended only to serve for molding the gold tray. The gold tray was then held in place against the corner of the tooth, and fresh oxyphosphate inserted. When this was hardened the whole was finished with sand-paper disks, and presented the appearance, from the palatal aspect, of any ordi-

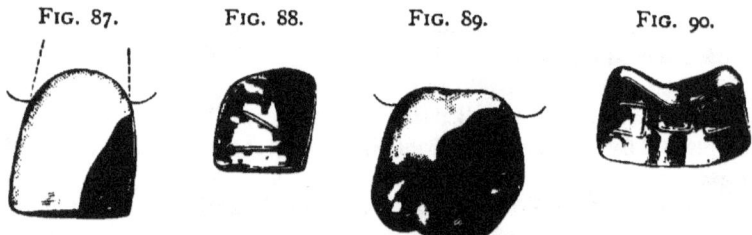

FIG. 87. FIG. 88. FIG. 89. FIG. 90.

nary gold filling, while labially I had the oxyphosphate simulating the tooth in color. In Fig. 87 is shown a tooth prepared for filling, the gold tray in position to receive the phosphate. Fig. 88 shows the tray itself, and the inner surface here represented with loops. One or two pins from old porcelain teeth, soldered into the tray, serve as well as any other means of obtaining a point for the phosphate to adhere against.

Another condition where extensive contours may be safely effected mainly with oxyphosphate, is where a large portion of a tooth being missing, let us suppose that we find the tooth itself intensely sensitive, so that we should hesitate to insert a metallic filling. Or the patient may be of such temperament that it would be hazardous or injudicious to compel as long a sitting as would be necessary for the insertion of gold. It may be desirable to reject amalgam, because the location is such that it would show, as for example the anterior approximal surface of a superior first molar, or one of the bicuspids. The plan that I have followed with success is to temporarily restore the shape of the tooth by inserting an oxyphosphate filling. Next I take an impression and dismiss the patient. During his absence I mold pure gold plate over the plaster model, forming a contoured cap to cover the oxyphosphate, which I insert at the next sitting. The inner side of this cap is treated as before, either with pins or loops soldered in. Fig. 89 shows a molar thus filled, the appearance being simply that of a gold contour filling, because the edges are polished down to the finest taper, which can be done so nicely that the point of an explorer will pass over them, and silk not catch under the gingival margin. Fig. 90 shows the cap with loops within.

It may be as well mentioned here as elsewhere, though not strictly speaking in the nature of contour work,—though all restoration may be so counted,—that these caps for oxyphosphate become invaluable in the treatment of *children's* permanent teeth. When the little ones come to us with gaping cavities in their sixth-year molars, what are we to do? We frequently find extensive caries without real exposure of the pulp. It is greatly desirable to save these teeth alive, as a dead pulp in a sixth-year molar, at this time, even though thoroughly removed, may almost surely be counted a forerunner of an abscess later in life, and

FIG. 91.   FIG. 92.

probable loss of the tooth before the age of twenty-five. So far from considering these teeth good subjects for extraction, I take unusual pains to save them. I think this can be accomplished with oxyphosphate better than with anything else. Here again I fill the cavity temporarily, in this condition placing cotton, which carries a medicament first, and covering with gutta-percha or temporary stopping, carving the same into fair shape for mastication. I then take an impression, and if it be necessary make dies and swage a piece of pure gold to cap my cavity and give me a good masticating surface. Fig. 91 shows a tooth so filled, and Fig. 92 the cap with loop. These caps should *never* be made to do service over *gutta-percha*, as that material by swelling may dislodge the gold covering. In all cases the edges should be as thoroughly polished down as though it were a true gold filling.

The Use of Screws.—The judicious use of screws in connection with the restoration of teeth which have been badly decayed, enables us to save permanently many teeth which would otherwise be lost—or crowned. The screw first comes into play when caries has advanced so far that any preparation of the cavity, removing tooth-substance, in order to obtain retentive shape, would leave the remaining portion so weakened that fracture subsequent to or during the filling operation would be rendered probable. Again we resort to the screw when necessary excavation might too nearly approach, or even expose the pulp itself. A case which combines both of these possibilities is shown in Fig. 93, which is a lateral incisor from which both corners have been removed by carious action. Supposing that the pulp were alive in such a case, I think it better to restore by gold contours than to crown, even where crowning could be accomplished without destruction of the pulp. Using the How drills, taps, and gold screws, a threaded hole is drilled in the tooth, care being taken neither to enter the pulp-chamber, nor to pass through the side. The screw is then turned into place tightly and cut off long enough to reach almost to the end of the corner which is to be builded on. The figure shows the screws in position. To secure permanent results, the greatest skill and care

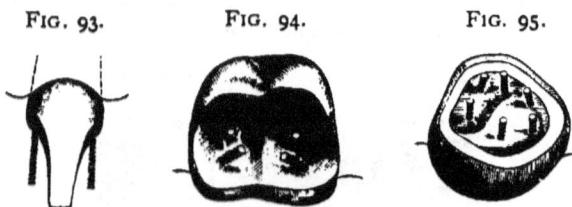

Fig. 93.   Fig. 94.   Fig. 95.

are requisite in filling around the screw. Only very small pieces should be used, and each of these should be thoroughly condensed. As a screw renders an operation doubly difficult, it should be resorted to only in cases of absolute necessity. Fig. 94 shows a good arrangement of screws in a molar, the pulp being alive, and the tooth so sensitive that proper excavation of the cavity becomes of doubtful expediency. Here we have four screws so placed that they assist one another. Where it is designed to depend upon amalgam, a screw of platinum and iridium is preferable to one of gold, which could be readily destroyed by amalgamation. We sometimes find molars badly scooped out by abrasion, a considerable surface of the dentine being thus denuded. Occasionally these places are extremely sensitive. Here screws may be depended upon to do good service. I have placed as many as six in a single tooth, as shown in Fig. 95, which represents a molar ready to receive the amalgam, no undercutting having been attempted, reliance being placed entirely upon the screws.

In concluding this subject, I will give a few odd cases from practice which may indicate the unusual application of principles which become necessary in unique cases.

Case 1. A young lady visiting the city called upon me with the statement that she had fallen and broken a lateral incisor. Her dentist, in the city where she resided, did not care to undertake the operation, and as she was about to visit New York kindly gave her my address. Examination showed that the tooth, if a lateral incisor at all, was misshapen, being indeed a supernumerary tooth. Though narrow, its loss would have been deplorable, because the arch was well curved and filled with teeth regularly placed. The fracture had removed about one-quarter of the crown, yet the pulp had not become exposed thereby. The end of the tooth was broken off square, and being small in proportion to the length of the restoration required, made the chances of success next to impossible if any attempt were made to form a cavity by undercutting, which should retain the filling without danger to the pulp. I built down a full contour, retaining it entirely by three screws. The arrangement before filling is shown in Fig. 96, and the restored end in Fig. 97. To start the filling, I took a narrow rope of gold and wove it between and around the screws.

FIG. 96.   FIG. 97.   FIG. 98.   FIG. 99.

Case 2. This case exemplifies another use of a screw. The patient was a young man, and the tooth a first bicuspid. Approximal cavities had approached each other till the cavity extended through from front to back, as seen in Fig. 98, which also shows the disposition which I made of a screw. The labial and palatal walls were so weak that they could be sprung together slightly between the fingers, so that I feared fracture from the force of mastication. As an additional precaution, the edges of the natural cusps were slightly beveled, so that after filling they were capped and protected by the gold, as shown in Fig. 99. The ends of the screw were ground off after filling, appearing like small fillings.

Case 3. A young woman of twenty, of comely features, applied to me to have a gold crown placed over a first bicuspid. Examination showed very extensive decay along the posterior approximal portion, which also extended so far up under the gum-margin that not only would it have been difficult, but I thought that it would even have been impossible to successfully crown the root after cutting off what was left of the natural crown. The condition is shown in Fig. 100, where it is seen that the cavity involves the root as well as the crown. The natural crown itself offers poor support for a filling, yet this tooth was successfully contoured, and is doing good service now, seven years later. It is manifest that the rubber-dam could not be forced above the gingival margin. The procedure therefore was to secure a stout platinum and iridium screw into the root-canal, the pulp having been removed before I saw the case. Around this, amalgam was packed until it protruded sufficiently to allow the dam to be placed. At the next sitting the contour was completed with gold, the dam being placed first, and then all the amalgam cut away except just sufficient to allow the dam to remain in position. Fig. 101 is a section showing the relation between screw, amalgam, and gold.

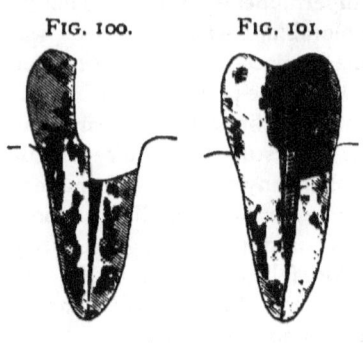

FIG. 100.   FIG. 101.

Case 4. A young man came to me on one occasion with a first superior molar from which the crown was lost except the buccal wall, yet the pulp was alive. A good, serviceable restoration was made without cutting away the tooth to any material extent. Along the palatal margin two iridiumized platinum screws were placed, and through the buccal wall itself two holes were drilled. This wall was strong, and yet of such form that it was safer to resort to this method than it would have been to make an attempt to undercut it for retaining grooves. Fig. 102 shows the tooth as prepared for filling. These holes were beveled at the buccal side so that when the amalgam was placed it extended through, forming headed pins, which materially added to the strength of the operation.

FIG. 102.

Case 5. A married woman of forty came to me for a lower set of teeth. She still had the six anterior teeth below, but they were in such a state of dilapidation that she wished them extracted. Examination showed that though badly decayed along anterior and posterior approximal sides, in every instance the pulp was alive. For this reason, and because I considered that a partial lower plate would give better service than a full denture, I advised saving the natural teeth. In every case, I found the teeth so narrow and the cavities so extensive that I despaired of being able to build on the necessary corners, with any hope of having them endure. Finally I resorted to a plan which proved successful. One case will serve for illustration. Fig. 103 shows a lower incisor after the removal of decay. It is seen to be similar to the superior lateral incisor shown in Fig. 93, where screws were resorted to. This could not be done in the lower because of the danger of entering the pulp-chamber. What I did was to connect the cavities above and below, cutting a groove across the incisive edge, and another near the gum at the labial aspect. The cavity prepared for filling is shown in Fig. 104, and the full contour in Fig. 105.

Case 6. A young girl of about fourteen presented with a lower sixth-year molar, from which all of the crown was absent, the pulps, however, being still alive. These cases are by no means uncommon. Of course, where the patients can afford it, they may be crowned with gold, but even then it is doubtful whether such an operation would be as serviceable as the one which I am about to describe, for two reasons : First, the fact that by the method of filling *the dam is in position when the cement is used;* and, secondly, a better occlusion

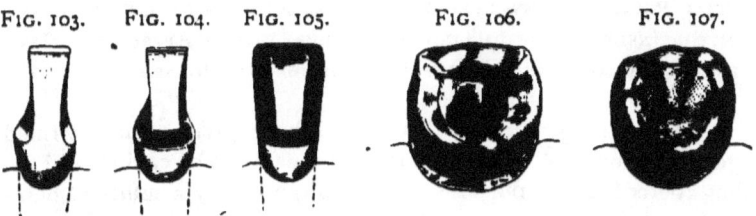

FIG. 103.   FIG. 104.   FIG. 105.   FIG. 106.   FIG. 107.

can be obtained *where the "bite" is short.* I made a band of gold to fit around the tooth, wide enough so that when in place it prevented closure of the jaws. It was then removed, and the upper edge turned inward slightly all around. Replaced about the tooth, closure of the jaws pressed the bent edges down until the occlusion was perfect. The tooth with band thus turned in to adapt itself to the occlusion is shown in Fig. 106. The dam was next adjusted, and cement placed along the inner side of the band and covering the floor of the cavity thus formed. Before this was allowed to set, amalgam was used to complete the contour. This cement not only served to secure the superstructure to the root, but it protected the gold band from the action of the mercury in the amalgam. When trimmed up and polished, a handsome and serviceable crown resulted, as shown in Fig. 107.

FIG. 108.   FIG. 109.

Case 7. A young woman presented with a first bicuspid from which the entire labial portion had been lost. The tooth was pulpless. I ground a porcelain cuspid to fit fairly well into the place of the missing labial wall, and to this, when backed, soldered a stout platinum bar, which extended down into the root. This was set with oxyphos-

phate and the patient dismissed. At the next sitting the dam was placed, and the phosphate removed sufficiently to allow a thorough filling of gold. Next a groove was cut between the porcelain face and the tooth where they came into contact at the labial gum-margin, and this joint was made perfect by gold filling. Fig. 108 shows a section representing the relation between porcelain face, pin, phosphate, and gold, and Fig. 109 shows the appearance from the labial aspect, the porcelain face, and the gold at the joint being seen.

CHAPTER V.

Special Principles Involved in the Preparation of Cavities, and in the Insertion of Fillings—Consideration of Approximal Cavities in Incisors—In Cuspids—In Bicuspids—In Molars.

In the earlier portion of this work I have given what I termed "general principles" in relation to the preparation of cavities and the filling thereof. I now purpose to take up specifically a sufficient number of typical conditions, so that with the description of methods necessary, the student may have the theoretical knowledge which will enable him intelligently to undertake whatever may come into his hands.

In the first chapter I classified cavities as being of three kinds, viz : approximal, crown, and surface, the latter including palatal, lingual, labial, buccal, and festoon cavities. Of these terms, all serve well enough, with the possible exception of the word "crown." Strictly speaking a tooth is divided into crown, neck, and root, so that *any* cavity might be considered a "crown cavity" provided it did not reach or pass the neck. But this use of the word is arbitrary, for the common meaning of "crown" is "the topmost part," "the summit." Thus the "crown of the head" is the extreme upper surface. This latter application of the term has been made by many when speaking of cavities in the masticating surfaces of molars and bicuspids. In the absence of a universally adopted term applicable to this position, the words "crown cavity" in this work must be interpreted to mean a cavity in the masticating surface of a bicuspid or a molar.

Approximal Cavities.—Of all cavities, those in the approximal surfaces usually demand the greatest skill and care. Situated most often in a position inaccessible because of the contiguity of the adjacent tooth, a spreading, or separation of the teeth, becomes necessary. Even after this is done, it not infrequently occurs that what would be an easy filling with ready access, becomes exceedingly difficult because of lack of space in which to work. Add to this the tenderness caused by *any* system of wedging, and approximal cavities become, of all, the most to be dreaded by patient and operator. They usually have their initiation at the point of contact. Two teeth touch each other at a limited area, and caries begins at that spot. After spreading the teeth, and removing stains, there is seen at this point of contact a whitish, or chalky, appearance to the enamel. Except in its most incipient form, it is futile to attempt to remove this by filing or by polishing. The caries usually penetrates very deep in proportion to its circumferential area, and the dentine is swiftly reached. To eradicate the caries with the file or disk would be to expose the dentine, and at all events to destroy the integrity of the abutments of the arch by producing a space, which is almost always mischievous. The best plan is to accept the condition, as a cavity needing a gold filling. I am speaking of permanent teeth, and advise gold in all teeth at all ages, where these minute cavities are discovered.

Fig. 110 shows such a cavity in a central incisor. It should be prepared with as little extension of the borders as possible, although mak-

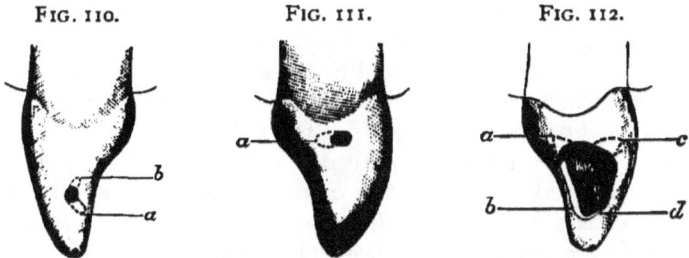

FIG. 110.   FIG. 111.   FIG. 112.

ing it a point to remove all carious material until strong edges are reached. There is a temptation in these cases toward enlargement, because of the fact that by so doing the work is rendered more easy. The excuse, however, is insufficient, and the practice reprehensible. Indeed, with me it has been my pride to make perfect fillings of the most minute kinds, and under the most trying circumstances. To retain the filling, a rose bur may be used, dipping slightly toward the gingiva as indicated at *b* and similarly toward the incisive edge as at *a*. Here at once we find a difference between "general principles" and "special cases." I have in a former chapter argued that no undercutting should be toward the incisive edge, or at least that it should not be extensive. But in these tiny cavities, where there is strong tooth-substance in all directions, we may allow ourselves the most convenient method and arrange the retaining points opposite to each other as indicated. Even in these tiny cavities I require two extensions of the cavity for good retention; yet again this is a rule with an exception. Should the cavity occur nearer the neck of the tooth, in the bulbous portion of incisors, or in a cuspid or bicuspid, in either of which there is considerable tooth-substance laterally from labial to lingual aspects, and should there be very limited space in which to work, it may often be both permissible and advantageous to form the cavity as a single deep pit. Fig. 111 shows such a condition in a cuspid, the dotted line *a* showing the inner extension toward the palatal aspect.

In filling these cavities with gold, I follow strictly the rule previously advanced, to place in a cavity only pellets which will pass the orifice without compression. I therefore prepare special pellets, tiny in size, and use sometimes as many as a dozen or more for a filling which when finished is no larger than the head of a small pin. Thus I am satisfied that I obtain as perfect a filling as when the cavity is larger and therefore easier.

Fig. 112 shows the opposite extreme, and we have the largest approximal cavity possible in a central incisor, without encroaching on either the palatal or labial surface. The preparation of this and of all cavities between it and the one in Fig. 110 comes immediately under the rule as already described in connection with Figs. 2 to 8 inclusive. To fill with gold, the first pellet should be placed in the palato-gingival extension at *a*, and should be fixed so firmly that the second and all subsequent pellets may be added without tipping out. The filling should be first extended along the palatal groove in the direction of *b*, great care being observed that this palatal edge shall be perfectly covered. Next, gold is carried toward and into the labio-gingival extension at *c*, after which the completion follows naturally, the labial edge being covered last, and the last pellet of gold being placed at *d*. If this be analyzed, it will be found that I have here followed the general and most valuable rule, "*Fill that part of the cavity first which is farthest from you.*" If we may fill teeth by rule at all, I should say that this one axiom has been of greater benefit to me in practice than any other in the whole realm of dentistry. Its no less important corollary is, "*Where two approximal cavities, adjacent to each other, are to be filled, fill that one first which would be least accessible were the other tooth non-carious.*"

In Fig. 113 we see a large approximal cavity which encroaches upon the labial surface. Under ordinary circumstances this should give the operator little or no trouble. In addition to the space obtained by the wedging, the fact that a part of the labial surface is absent furnishes an abundance of space in which to work. Yet the mistake should not be made that the cavity itself affords sufficient space without wedging. In a few instances this may be true, but more frequently a filling so placed when completed, though it may look perfect to the eye of the patient, would not prove to be so if examined by an expert. The palatal border would show defects, and it is because gold must be builded over this palatal border that more space must be obtained than is furnished by the loss of tooth-substance. To arrange such a cavity so as to retain a filling is readily accomplished, except where the pulp is nearly approached, a condition which renders more difficult the preparation of any form of cavity. Deep extensions are to be made at *a* and *b*, the labio-gingival and the palato-gingival angles of the cavity. These, where it is possible, should be deep enough to hold the filling of themselves. A slight groove should extend along the palatal portion *c*, but care should be observed at this point. It should be neither too deep nor too near the pulp, nor too near the edge of the cavity. Its object is not so much to add to the retentive strength of the cavity as to facilitate the packing of the gold, when the points at *a* and *b* having been filled and connected, we come to extend the gold along this portion. If there be a slight groove we avoid tipping. At *d*, which is toward the incisive edge, we avoid anything in the nature of an undercut, but it is wise to produce a well-marked concavity which will serve as a counterpoise to the upper retainers. Along the labial border *e* there should be no grooving, or undercutting of any kind. Such a procedure only undermines the enamel, producing a weak edge with probable fracture during the operation, or at least the production of a crack which will later bring the tooth back to us with an imperfection at this point. No strength whatever is gained for the filling, so that nothing but mischief can accrue.

To fill this cavity with gold, the first pellet is to be placed in the palato-gingival extension *b*, which should be so formed that it will hold it without tipping. I anneal this pellet and mallet it to place. It should be large enough to be readily wedged to place. The succeeding pellets should be small enough to reach their predecessors without compression, and each should be malleted thoroughly. The correct shaping of a retaining point is rendered futile unless the filling be solidly packed into it. Otherwise, though it may retain what is in it, it will fail to retain the bulk of the filling to which it should lend strength. From the palato-gingival extension proceed toward the labio-gingival pit *a*, and when both are thoroughly and solidly filled, proceed to cover the palatal groove *c* and the cavity-edge at this point. Here it will be wise to use hand-pressure for a brief period. With this milder force pack the gold along the groove and over the edge until a thin layer reaches the incisive concavity at *d*, when the mallet may be resumed. In this way we avoid fracturing the palatal edge by saving it from direct contact with the mallet-stroke. From this point on, the only care needed is to constantly bear in mind the rule, to fill the part farthest from us first, being sure that it is made full enough. Otherwise we find the filling pitted when we come to polish it.

Fig. 114 is the direct opposite of the last, and yet there are points of similarity which some do not note. The approximal cavity here encroaches upon the palatal surface. Where the loss of the labial surface simplifies the operation, the absence of a part of the palatal surface greatly complicates it. There is a curious fact to be observed here. Whereas many look upon the cavity shown in Fig. 113 as one which easily retains a filling, some are much disturbed by the cavity in Fig. 114. The first, apparently, is safe, because a filling will not jump *up* toward the labial aspect, whilst, *per contra*, gravitation may have a tendency to cause a filling to drop *down* and out at the palatal opening. Those who argue thus are deceived into this error of judgment by the fact that when the cavity is seen the patient is in the recumbent position, which makes the labial side of a tooth *up*, and the palatal side *down*. It should be remembered that when the patient is standing erect there is neither up nor down to this condition, and therefore whatever arrangement will retain a filling in a labio-approx-

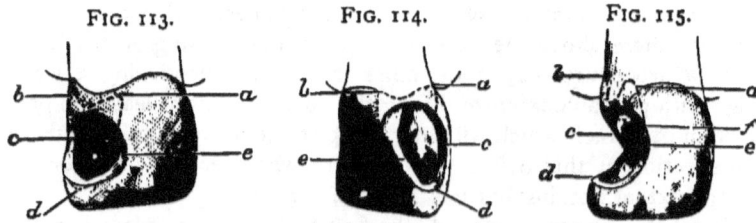

Fig. 113. Fig. 114. Fig. 115.

imal cavity will serve equally well in a palato-approximal one. Consequently the formation of the two classes of cavity are similar, though not identical. We have the same extensions at *a* and *b*, the same groove at *c*, the same concavity at *d*, and the same absence of undercutting at *e*. But we find that *c* and *e* have changed places in the two illustrations. In the previous one *c* represented the palatal part, whereas now it becomes the labial. A similar change occurs in placing the gold. The first pellet is to be placed in the labio-gingival extension, and, after the palato-gingival retainer has been connected with it, we then follow along the groove, which here is at the labial instead of the palatal border, so that we practically reverse the order of procedure. Yet really we do not, for now we are using a mirror, and in the reflection this labial groove, though actually nearer to us, *appears* to be farthest from us, and in this instance we must fill by appearances. If the operator is not skilled enough to use the mirror, he then tips the chair back and lowers his head, so that in that position the labial groove becomes really the farthest point from his vision.

In Fig. 115 we see a cavity which is practically a combination of the last two. We have presented an approximal cavity in a central incisor which encroaches upon both the labial and the palatal surfaces. Where these cavities are extensive, as pictured, they are usually trying. The question has been raised by some authorities as to whether or not the whole corner should be removed. I have already sufficiently expressed my own views upon this. If in either of the previous two cases we gained anything by the presence of the labial or palatal wall which remained intact, what are we to do here in the absence of both? In each of the other cases I argued that the groove which I advocate is not for retention, but to facilitate filling. Consequently, in the present instance, when both walls are absent, we may still feel safe with precisely the same arrangement,—labio- and palato-gingival extensions at *a* and *b*, a slight concavity at *d*, and a groove as before. The question now arises, Shall this groove be placed at the labial or at the palatal border of the cavity? The answer indicates the method of proceeding with the filling, for the groove is only intended as a leading gutter for the gold as we pass from the gingival toward the incisive portion. In this instance we place the groove along the palatal border *c*, for that becomes the distant portion, and must be filled prior to the labial.

I have said that this arrangement will be safe, but it will be so only so long as the natural corner remains intact. Elsewhere I have said that the filling must be so strongly anchored above that, in case of subsequent loss of the natural corner, a gold substitute may be built on without removal of the original filling. To accomplish this the labio- and palato-gingival extensions must be made as deep as possible with safety, and must continue as deep grooves toward the incisive end, relatively about as far as *f* on the labial aspect and *c* at the palatal. The upper half of the filling is thus securely held in three directions, the palatal, the gingival, and the labial.

In filling, after connecting the upper retainers and building along the palatal border into the incisive concavity, care must be taken to restore the lost palatal wall before attempting to fill the main part of the cavity. In plainer language, manipulate the gold so that presently the cavity, partly filled, will appear similar to that in Fig. 113, where, the palatal wall being intact, I said the filling becomes simple and easy. If another method be pursued and the whole of the cavity be floored over first so that it is lined with gold, we but add to our work, for we simply reduce the *size* of a difficult cavity without altering its *shape*. When half completed, the continuance will be more difficult than the beginning, whereas in the method which I advise we constantly lessen the task, making the cavity simpler and simpler as *we continuously alter the shape*. But even by this method only the most skillful operator can have a perfect filling when he has placed the last pellet along the labial border *e*, so that the work seems ended. Examination of the palatal surface of the filling with a mirror will almost invariably disclose the fact that the young operator has not made the filling full enough, or the surface solid enough, to allow for perfect contour after polishing. It is therefore essential for beginners, and wise for older men, to examine this aspect of the filling before removing the dam. Usually gold must be added and contour filled out. In fact, it should be the rule to polish all fillings before removal of the dam, except where in special cases a sufficient reason appears for finishing the filling at a subsequent sitting.

Next we must consider labio-approximal and palato-approximal cavities of another class. Fig. 116 shows a labio-approximal cavity which differs from that seen in Fig. 113 in that the caries has removed the enamel as far as the cutting-edge, along the labial border. This is a rare condition, but is seen occasionally.

The preparation of this cavity is practically similar to that of Fig. 115. The filling must be retained from above. Here, however, it is

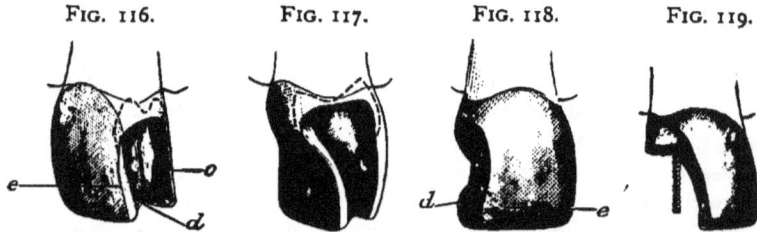

FIG. 116.   FIG. 117.   FIG. 118.   FIG. 119.

impossible to make an incisive concavity at *d*, because of the absence of the labial plate. It is undesirable to form an undercut at this point, and we should also abandon the groove along the palatal border below *c*, but it may be made along the labial border *e* with some advantage. Unless the palatal plate be excessively weak, it is seen from the foregoing that I should not remove it. It will serve as a protection to the filling during mastication, relieving it from the strain that would endanger the contour filling which would result were this part of the tooth sacrificed. At the same time the filling must protect it also, as well as be protected by it. To accomplish this, the extreme cutting-edge should be removed, as shown in the illustration, so that the gold may be built over it, thus protecting it from the antagonism of occlusion. No special directions are needed for the packing of the gold beyond the cautions already emphasized to pack solidly, using small pellets, and heavy foil for the final portion.

Fig. 117 shows a palato-approximal cavity similar to the last, in that the depredation reaches the cutting-edge. This is one of the most trying cavities that can be presented, and one, too, which occurs with unpleasant frequency. To remove the labial plate would be to simplify matters vastly, but if such a procedure was wrong in the last cavity, it would be doubly so in this. In the former instance the retention of the palatal plate gave only an added support to the gold, whereas here the labial plate not only serves the same purpose, but *it also covers and hides the gold from view.* An extensive filling of this nature may be placed and scarcely be seen from a front view. Wherever there is found sufficient strength, therefore, to this part of a tooth, I should allow it to remain. As in the last case, the cutting-edge should be removed, allowing the gold to be built over it, thus protecting it. But for cosmetic effects only the merest trifle should be taken away, so that barely a line of gold will show from the front. The formation of the cavity is the same as in the last instance, save that the groove now occurs along the palatal border, to avoid weakening the labial plate.

The insertion of the filling will tax the skill of beginners, as it does that of many of mature experience. This is one of those conditions where a judicious alternation of mallet and hand-pressure produces the most satisfactory results. Because of the inaccessibility, and because of the difficulty to see all parts of the cavity, the danger from fracture by the mallet blow is increased tenfold. My habit is to partly condense every piece by hand-pressure before taking up the mallet. Of course this is slow, but rapidity is not invariably preferable. The main point always is to obtain the best result, in safety. If the operator has a chair which will allow him to tip his patient back, so that he can see directly into the cavity, it would be a most advantageous procedure. If not, he must be skilled in the use of a mirror and depend upon it. After filling the labio- and palato-gingival extensions, I should gradually extend the gold toward the cutting-edge, following the groove and covering the whole inner surface with a veneer of gold. Up to this point I should use the mallet but sparingly, but with this protection supplied to the weak wall, the mallet may be used for completing the filling with less fear. A momentary warning, however, against too free a use of hand-pressure! Beware lest in the attempt to conscientiously condense the gold too much force be exerted, and a fracture occur. *More teeth have been broken by hand-pressure than with the mallet.*

Next we reach the condition where the depredation has removed the corner. The restoration of these teeth is classed with contour fillings, but they are also approximal cavities. An important warning is to be given at the very outset. Be sure to obtain sufficient space in which to work. The absence of the corner is apt to prove misleading. Apparently there may be abundance of space, especially where the cavity is extensive. In reality there never is, where the teeth are normally situated, and the neighbor has not been removed. The test will be at the gingiva. Here it will be seen at once that but little space exists between the teeth. Whilst the restoration may be made without separation, the polishing of the approximal surface of the filling must necessarily produce a permanent space. This can be avoided only by spreading the teeth, overbuilding the gold, and polishing so that when completed the original width and shape of the tooth is restored.

In Fig. 118 we see a central incisor from which the corner has been lost. Is any special direction needed for the proper preparation of the cavity? In speaking of the cavities illustrated by Figs. 113, 114, 116, and 117, I have said that the groove at the labial or at the palatal border was less for retentive purpose than for facilitating the operation of filling. In Fig. 115, however, I advised that the upper retaining extensions should terminate toward the incisive edge in grooves which extend about half-way. This was in view of the possible future loss of the corner. Thus it is seen that *though I do not depend upon lateral grooves in the presence of the natural corner, in its absence I do*, more or less. Thus in such a cavity as shown in Fig. 118, after arranging the upper portion as described in Fig. 115, I should continue the grooves toward the incisive region, lessening them in depth and extent until they meet about at *d*. Some dentists make a deep dip inward, that is to say toward the pulp, at this point, endeavoring to obtain a retaining-pit. The result usually is that because of the narrowness of the tooth in this region, this undercutting leaves the labial plate of enamel so thin that the gold is seen through it when the filling is completed. Worse than this, the enamel often appears cracked, and I have seen not a few cases where it has chipped out afterward. Let us consider for a moment what is gained by an extension such as is indicated by the dotted line *e*.

The object of any undercutting is of course to prevent the filling from being forced out of the cavity. Supposing, then, that a strain tends to press the filling out laterally, that is to say toward the adjacent tooth, will this incisive undercut prevent this? If the surrounding wall be quite strong it might have such a tendency, but the real resistance would be found in the palatal and labial grooves, which, *extending the full length of the filling*, thoroughly protect it. The filling will scarcely be movable toward the gingival wall, for there we have the greatest resistance and the best arrangement for retention. Can the filling move downward toward the incisive edge? If so, this undercut might serve. But there is nothing but gravitation to urge such a movement, and that is so slight that the upper extensions are a million times more than adequate. Consequently we find that nothing is gained by the procedure. Is anything lost? *All undercutting which is unnecessary to the retention of a filling is mischievous.* In this instance it would be especially so, since it weakens the incisive end of the tooth, the very point which must sustain the full and first strain of mastication.

It follows, then, that we may depend entirely upon the labio- and palato-gingival extensions, together with the labial and palatal grooving, to retain approximal contours such as shown in this figure.

When the depredation becomes so extensive that the grooves would encroach upon the living pulp, we are compelled to adopt new methods. Fig. 119 shows a cavity of this nature. Strictly speaking, the line of the original pulp-chamber has here been passed, the pulp having receded before the approach of the decay, a common occurrence. Sometimes we may even see the plain outline of what was once the chamber now filled with secondary dentine. To make the lateral grooves here would be hazardous. To arrange the cavity as described in Fig. 115, depending only upon the upper retainers, might prove ineffectual. In this dilemma the screw comes to our assistance, arranged as in the illustration. Up to the point where the screw becomes a necessity there is no imaginable cavity which cannot be shaped to retain a filling, and in teeth from which the pulps have been removed we are seldom driven even to this extremity. I have built down from the gum line complete crowns on centrals, laterals, cuspids, and bicuspids, without resorting to screw or post, and yet have obtained durable (though according to present standards and in the presence of the porcelain crown unsightly) results.

Where we use the screw to retain such a filling as would be needed in Fig. 119, we still have the labio- and palato-gingival extensions, the screw passing upward along the median line. My method of filling is to pack my gold solidly into my upper retainers first, connecting them. Then I drill through the gold thus placed and into the tooth-substance beyond. The drill-hole is then tapped, the screw turned tightly into place, and cut off just short enough not to reach to the line of the incisive edge, after which the filling is completed with small pieces, care being taken to properly surround the screw.

FIG. 120.

In Fig. 120 we have the extreme condition where but one corner is involved. These cases are usually the result of fracture, though they sometimes occur where such a cavity as Fig. 119 has been improperly filled, with the result of subsequent decay and destruction along the incisive edge. As the latter would be the simpler, I will speak only of such as result from accidental blows.

These cases are frequently very puzzling. They most often occur during childhood. When they do, unless the pulp be exposed, as unfortunately sometimes occurs, or unless the pulp should die as the result of the concussion, it is wiser not to attempt any operation at all until the sixteenth to eighteenth year. By that time, especially if the patient be a female, something must be done. Should it be decided to contour with gold, the first step will be to determine whether the two centrals may be shortened with impunity. This often is a material advantage. A case from practice will best illustrate. A patient came to me, a young woman, with two leaky fillings appearing as seen in Fig. 121. The two centrals drooped considerably below the laterals, as often occurs, and I shortened them. After refilling they appeared as in Fig. 122. To return to Fig. 120; should shortening not be desirable, which of course would reduce the condition to an approximation of Fig. 119, we must decide upon an arrangement of the cavity which will retain this unusually extensive contour. In the

FIG. 121.    FIG. 122.

main this will be the same as in Fig. 119, including the screw, but in addition to this, now that we may reach the opposite side of the pulp-canal we may with advantage form a retaining extension in the direction shown by the dotted line at *a*. It must not be forgotten, however, that there is danger of reaching an elongated cornua of the pulp at this point, and the preparation must be advanced with great caution, extreme sensitiveness being a danger signal not to be overlooked. Although this retainer would be an advantage, it is not a necessity, and should not be obtained at the expense of possible need for destruction of the pulp.

FIG. 123.

Up to this point it is of course immaterial whether the cavities described occur on the mesial or distal surface. In either case they would be treated alike. Should they occur on both surfaces of a tooth, each would be a condition by itself, without relation to the other. There comes a point, however, where approximal cavities occurring on both distal and mesial surfaces of the same tooth do bear a relation to one another, and materially alter the rule of management. Fig. 119 has been introduced to mark the point at which we abandon the sole dependence upon retentive shaping and resort to the screw. Should two such cavities occur in one tooth of nearly the pictured dimension, we would no longer need a screw. Fig. 123 shows such a tooth after the insertion of the filling. It suffices to convey my idea. It is seen that the slight portion of the incisive edge which remained standing has been entirely removed, so that the two cavities become thus united. By this procedure we have a single though extensive cavity to deal with, and the strong retaining powers of the gingival portion of the cavity at each side readily retain the whole mass, operating as they do to counterbalance each other. The cosmetic effect here is better than where an effort is made to preserve the trifling bit of the incisive edge of the natural tooth which remained, so that the arrangement is good both for durability and for appearance.

An analysis of the directions given up to this point discloses the fact that in every cavity the main reliance for retention is upon the depth and strength of the labio- and palato-gingival extensions. The latter especially is the dependence. In making these extensions we are of course obliged to avoid exposure of the pulp. Therefore they must be made between that organ and the external surfaces, and distant enough from each to avoid disaster. The pulp lies nearer to the labial surface at this part of the tooth than it does to the palatal, for the reason that on the palatal surface near the neck there is a considerable bulge. It is in this thick portion of the tooth that we can anchor many fillings which otherwise would seriously annoy us.

What, then, are we to do with those cavities which present with great depredation about the very part in which these two important anchorages are to be placed? A single example, choosing an extreme case, will serve for all. Fig. 124 shows a central incisor in a condition of almost total wreck. Caries has passed beyond the enamel line and encroached upon the root itself at the neck. It has burrowed along the labio-gingival angle until seemingly there is little hope for anchorage there. It has eaten away a good share of the palato-gingival angle, so that our mainstay in that position apparently must be abandoned. In such an extremity we might hope to depend upon lateral grooves, but here we find that tooth-structure has been lost until we are almost at the plane of the pulp-canal itself. Even what is left of the corner is weak. Yet the pulp is alive, and therefore if it can be filled with gold the tooth perhaps would last longer than were it fitted up with a crown. It can be filled, and well filled, with gold. How are we to proceed in this case? Give attention first to the labio-gingival angle, at *a*. The caries passes as a sort of horn toward the labial festoon. Use first a rose bur as wide as the cavity at *a*, and deepen as far as possible without danger to the pulp. Then with a small bur undercut this place toward the gum and toward the incisive edge. This forms what might be termed a dovetail. I have elsewhere said that I do not

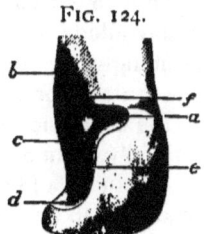

FIG. 124.

connect the labio-gingival and palato-gingival extensions by a groove. That is because ordinarily such undermining would weaken the enamel remaining at this point. Here we find that caries has removed the enamel. The root has been encroached upon. We need no longer dread fracture, since the enamel has vanished, and we may consequently make as deep a groove as possible, from our labial dovetail toward the palatal angle as shown at *f*. Now we reach the palato-gingival angle, or what is left of it. While we cannot get the strong retainer that we ordinarily obtain, yet the bulbous portion of the tooth extends all along the palatal aspect, and much of it is at our disposal. Continuing downward from the gingival groove, we make a deep undercut into this bulbous portion at *b*, and obtain an antagonizing retainer for the labial dovetail at *a*. Next with fine burs cut deep, narrow grooves along the labial and palatal borders, passing just between the pulp and the two outer surfaces, at *c* and *e*. At *d* attempt no more than the usual incisive concavity. I have filled just such cavities as this, where the caries had gone even farther and removed the natural corner. Consequently I know that even here we need no undercut at *d*. Lastly, with a small rose bur, dip the bur-head here and there, in all undercuts, and wherever it can be done with safety. The object is to make the surface of the cavity as rough as possible. If this cavity be filled with quite small pieces, made thoroughly cohesive, and packed carefully into every undercut and roughness, it will be found in place years afterward.

I believe I may now pass from the central incisor to other teeth. Of lateral incisors there is little to be said which is not covered by cases in the central. It must be remembered that being a smaller tooth it is frailer, and therefore very deep undercuts are to be decided upon with extreme caution. There is one very odd circumstance that I may allude to. I have noted, after a careful study of the facts, that seemingly the superior lateral incisor is more prone to abscess than the centrals, or possibly than any other teeth at all. I have also come to the conclusion that the pulp of a lateral incisor is less responsive, painfully, than any other. It is more easy to destroy, and more apt to die under a filling. In short, it seems to have less vitality. This leads one to be more cautious in preparing cavities, for it is not at all uncommon to find a pulp exposed after one has begun to place his gold, when the excessive heat of a freshly annealed pellet at last causing pain attracts the dentist's closer attention, and causes him to discover a minute exposure, over which engine-burs and excavators may have passed without remonstrance from the patient. Thus we might easily, through rapid excavation, actually expose a pulp, which accident could have been avoided by more caution, born of the knowledge that lateral incisor pulps are comparatively irresponsive.

In the cuspids we have the same general rules, with one or two notable exceptions. I have argued against any undercutting at the incisive corner of a cavity, and also to some extent against extensive grooving along the labial and palatal borders. This is changed in the cuspid, for here we have proportionately more tooth-substance between the pulp and the outer surfaces. Fig. 125 shows a cuspid having a cavity similar to that in the central in Fig. 115. What was a quite difficult condition in the central is much more simple in the cuspid, so far as preparation of the cavity is concerned. Where it is the distal surface which needs attention, of course the actual filling is more difficult. To form this cavity in the cuspid, we need not make the labio-gingival extension as deep as before, but as there is abundance of tooth-substance at the palato-gingival angle we may, if we desire, make the extension even deeper, though it is rarely necessary. Again, as discretion may dictate, we here may make extensive grooves along

FIG. 125.   FIG. 126.   FIG. 127.

both labial and palatal borders, and quite a deep undercut at the incisive corner *d* is permissible where requisite. This last departure from the former rigid rule is allowable, first, because of the excess of dentine in the cuspid beyond that found in the incisors, and second, because the pointed cusp of the cuspid renders fracture of the corner much less liable to occur than in the incisor, which presents a broad, thin incising edge to combat the strain of mastication.

In Fig. 126 we see a cuspid needing a contour filling. When I state that I have never seen a cuspid in which I found it requisite to use a screw, it is seen at once that this tooth offers a better field for extensive operations than the incisors. In Fig. 126 the full corner may be safely restored, depending entirely upon the internal arrangements. We make both the labio- and palato-gingival extensions deep. We unite these with a deep groove, for in this tooth we may feel comparatively safe against that fracture of the gingival enamel, which makes the procedure prohibited in incisors. We make strong grooves along both labial and palatal borders, and we make exactly the undercut at *e* which was forbidden in Fig. 118. Indeed, this undercut in the cuspids is mainly the reliance which makes us feel safe without the use of a screw, and it is therefore resorted to in all cases of contour. Thus we find that the cuspid is comparatively easy to work upon. Distal cavities are often troublesome, especially when the condition is what I have called a palato-approximal cavity. If it is found difficult to obtain much separation, which is not at all unusual, the work is sure to be very tedious. Yet even farther back than the cuspid region distal cavities may be made to seem almost like mesial, if the patient be tilted sufficiently to bring the cavity within range of the eye, as is often possible.

There is one cavity which we sometimes find in a cuspid, to which I may make special allusion. It is shown in Fig. 127, where it is seen to occupy a position along the neck of the tooth just above the bulge of the enamel, and extending around toward the palatal portion. This kind of cavity is most often found where an ill-fitting clasp has been used to retain an artificial denture. Where this is the case, the first bicuspid is commonly absent, which facilitates the operation. But let us suppose that the first bicuspid is not absent, and that the clasp has stupidly been placed between the two teeth, encircling the cuspid. To produce retentive shape, all that is needed is to cleanse out all decay and then undercut along the full extent toward the gum and toward the incisive edge. At the palatal end of the cavity, b, make a pit deep enough to hold the first pellet of gold securely; in filling, use the mirror, and with hand-pressure build the gold from that point well around into the approximal portion. It may then be thoroughly condensed with the mallet, and the filling completed as though only the approximal portion had been involved. Great care is needed to properly polish this filling, because it is above the bulge of enamel.

I may pass now to bicuspids, and at once we approach an almost entirely different field. Many dentists claim that the approximal cavity in a bicuspid, especially in the distal surface, is the most trying in the mouth. Omitting extreme or unusual cases, this is probably true. An evidence of this is the fact that recurrence of decay at the gingival border of fillings is more often found in bicuspids than elsewhere, with the possible exception of molars, which are much the same. This fact is also an argument that such recurrence is rather due to faulty filling than to any idiopathic causes connected with the position, as is claimed by many.

In general, the main difficulty is that it is harder to obtain sufficient space. This is shown readily, for in the absence of the adjacent tooth, even the distal approximal cavity in a bicuspid is really simpler than a similar cavity in an incisor. That this should be so is due to the fact that the bicuspid is a wide tooth, and through a narrow space it is often quite difficult to reach all parts of the cavity. When the bicuspid happens to be unusually long, and the cavity nevertheless reaches nearly to the gum line, the same difficulty is manifestly increased. I have seen teeth of this nature which persistently resisted all efforts to obtain much space, and which, containing extremely large but shallow cavities, became excessively tedious in the filling, because only very small pieces could be properly carried to place. Yet we are told that a steel separator is sufficient for all purposes, and again that a matrix should be placed around all approximal cavities in bicuspids and in molars! I must give a few illustrations to show the main differences in preparation of cavities, from what has been already said.

In Fig. 128 we see a bicuspid in which I have indicated two small cavities. It is evident that either of these would be simplicity itself in the absence of the adjacent tooth, in which case abundance of space would give ready access. In an ordinary tooth where a fair separation is procurable, either would still be manageable. With inadequate space, and in a wide tooth, either of these will harass both patient and operator if the attempt be made to prepare and fill without extension of the cavity. A pellet of gold crowded between the teeth will be so compressed that by the time it is forced into the cavity itself it is unfit for use, quickly balling up so that it must be discarded. Again and again the same mishap may occur, until it becomes easy to lose patience at the difficulty of filling what from its shape alone should be so simple. But with inadequate space no cavity is simple, and I might almost say the smaller it is the more annoying it will be. Therefore, in such cases I should recommend extension of the cavity to make it accessible. The direction of this extension depends upon the position. It should be toward the labial surface, or the crown, as it happens to be nearer the one or the other. For this reason the illustration shows two cavities, the position of one of which makes extension toward the labial surface preferable, as shown by the dotted lines, whilst in the other I would cut through to the crown. The first practice will be resorted to less frequently in the mesial than in the distal surface, because of the probable showing of the filling. To cut through to the crown will be chosen oftener on either approximal surface. I must pause here a moment to defend this proposition. It has been argued by a skillful operator, and careful thinker and writer, that the crown of a bicuspid or molar should under no circumstances be disturbed in this way. An allusion is made to the similarity here to the strength of an arch. That is, the portion of the tooth which I advise removing, this writer claims is as the arch to the cavity, and its greatest protection. A better argument that he uses is, that it assists in holding together the labial and palatal plates. All of this is beautiful theory; but to my mind it is illusory, if not false. Given a

FIG. 128.

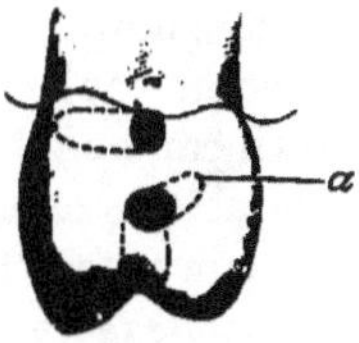

bicuspid or a molar, sound in other respects, but having such a cavity as the one in Fig. 128 which is nearer to the crown, and the removal of enough tooth-substance to reach the cavity from the crown, as is indicated by the dotted lines, leaves us a tooth with which one could eat for years without fear of fracture *even if it were left unfilled*, imagining for the sake of argument that caries would not supervene. This being, as it is, true, it follows that when protected by a perfect gold filling the tooth will be safe enough. In the last statement I based my argument upon mechanical laws, for the moment forgetting physiological ones, yet it is by physiological laws that the theory of not cutting through to the crown is most easily proven to be fallacious. The whole aim of a filling is to restore the depredations of caries and to prevent a recurrence. Granting now for a moment that the cavity is better made without cutting to the crown, which of course is true, yet if it cannot be made fully accessible by separating, it will be improperly filled, so that caries, which is not hindered by mechanics, will be sure to recur, and destroy that arch which the operator so thoughtfully preserved. Of course there are men so skillful and so patient that they can make perfect fillings through the narrowest of spaces. But these men are rare, and the best set of principles is that which may be employed by the greatest number.

Now, that I may not be misapprehended, let me state my position once more, succinctly. The borders of a cavity should never be extended when a dentist can by a fair expenditure of time and skill make a perfect filling. However, wherever from lack of space, or other good cause, a perfect filling can be better made by extension of the borders, it may be done, and such extension may be allowed to encroach upon the crown. Indeed, there are cases where there will be no other adequate method of properly retaining a filling, as has been already shown.

Another point is pertinent here. Supposing that such cavities as shown can be filled with amalgam without extension, would that be preferable to extending the borders and filling with gold? I should decide in favor of the gold filling in every instance. *Gold is always preferable to amalgam for durability, wherever it can be inserted producing what we term a perfect filling.* Dogmatic, but true.

The retentive shaping of these two cavities differs somewhat. In the one nearer the crown, very little undercutting should be made along the labial border, but a considerable groove may be cut along the palatal, with a distinct dip toward the gingival corner as indicated by the dotted line *a*. In the other, which after extension will be filled through its labial opening, a moderate groove toward the gum and one toward the crown will suffice. At the palatal end, here the bottom of the cavity, a depression may be made deep enough to retain the first pellet immovable. Either of these cavities, prepared as indicated, becomes simple, and could be rapidly filled with gold.

In Fig. 129 we have what may be termed a simple approximal cavity in a bicuspid. It is similar to Fig. 112, where it is shown in a central. Its preparation here is different. We no longer need deep labio- and palato-gingival extensions, nor are we compelled to avoid labial and palatal grooving. In the bicuspid we have simply a cavity surrounded by strong walls. Any antagonizing retaining arrangement will serve to hold the filling in place. The choice, therefore, must depend upon attendant circumstances. Where the adjacent tooth is missing, a moderate groove forming an undercut all around will prove the simplest and best. Where there is an adjacent tooth and we must work at the disadvantage of having the cavity more or less inaccessible, we may best proceed otherwise. The most universally useful arrangement will be as follows : make a palato-gingival extension, not too deep, but shaped so as to restrain the first pellet, as indicated at *a*. Make a similar depression toward the crown at the palatal corner, as seen at *b*. Connect the two by a fairly deep groove along the palatal border. A rose bur run around the rest of the cavity borders, making a general but not deep groove, gives us a strong cavity readily filled. To place the gold, begin in the palato-gingival extension, and having securely anchored the gold at that point, build along the palatal groove and also along the gingival border simultaneously, being careful not to advance either too rapidly, lest the excess of gold placed at one point make the other inaccessible. This is a very important factor in the successful filling of a bicuspid. It will be seen that by following this plan we finally come to a point where the labial corner at *c* must receive attention. Especially where the labial border of the cavity is at some distance from the labial surface proper, any undercutting at this part becomes quite difficult to fill, and the most unmanageable point of all would be at *c*, so that as we approach this place we must use greater caution. It must be filled before the gold has been brought so close that we have inadequate room to reach it ; and yet in packing gold into this corner, if we at once build it out to the edge of the cavity, we will find it difficult to fill the hollow which will have been left behind along the surface of the gold. More than likely when we come to polish the filling we will find an ugly pit, caused by an insufficiency of gold. The remedy is simple, but requires patience. Though for the bulk of the filling we may have been using fairly large pellets, as soon as we arrive at this corner we must choose smaller pieces of gold and build

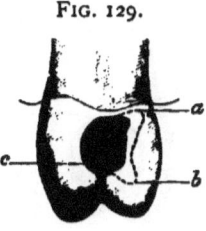

FIG. 129.

slowly. Strictly speaking, we should have been using heavy foil only, for some time previous to reaching this place. To fill the actual corner we should select small pellets, and as soon as these have brought the filling about half-way from the bottom of the cavity to the margin, we should return to heavy foil, which because of its form is peculiarly adapted to our needs here. With it, by proper manipulation, a filling should result which would be without a flaw.

FIG. 130.

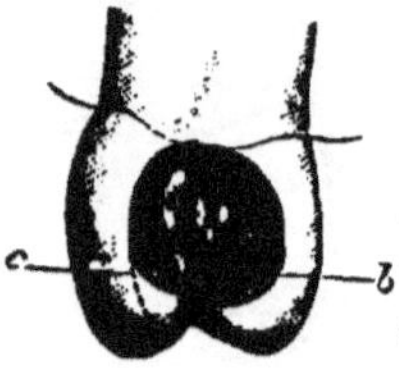

Fig. 130 shows a common form of cavity, such as is sometimes termed a compound cavity, because it occupies both crown and approximal surfaces. The arrangement for retention is precisely similar to the last, except that, as we now have an anchorage in the crown, we need not make the deep undercut at b. As before, c becomes a point of interest, greater here than in the last condition. If left as pictured, with a narrow passage connecting approximal and crown, the resulting sharp corner will give considerable trouble. The similar angle on the palatal side will not be as bad, though even there it may be trimmed down with advantage. At c, however, the labial angle should be removed as indicated by the dotted line, which, if studied, will be seen to follow the natural curve of the labial border line. When filling this cavity, as in the last case, this part will need extra care. Supposing that all has advanced as in the last, the best plan will be to give attention to the part of the cavity occupying the crown, and fill that thoroughly without special reference to the approximal portion. This done, we have reduced the case to exactly what we had in the last cavity when we came to fill c, and consequently can now proceed as then. Should we attempt to fill c before attending to the crown, we would find it very troublesome. In fact, it would be necessary to alternate between the approximal and the crown, causing much annoyance, especially as it often happens that the position of the patient must be altered so as better to see one or the other. To fill the crown first, simplifies the operation, and is therefore preferable. In these extensive approximal cavities, the gingival border is always a most trying obstacle to success. To obtain a perfect margin to these cases is as difficult as it is essential. I rarely make a deep groove or undercut here, but rather aim to leave the margin strong and very slightly beveled. More strictly speaking, instead of beveled —a word usually applied to the extreme edge of a cavity—I should say that I form it continuous with the floor of the cavity. This allows the placing of a fairly large pellet of gold, which, attached to that already secured in the palato-gingival extension, may be laid against and lapped over this gingival margin, and partly condensed by hand pressure, using a foot-plugger. Next I cover this with a piece of No. 60 gold and mallet it down, again using the foot instrument. By this method the whole border is well and rapidly covered, the under pellet serving as a soft cushion which easily adapts itself to irregularities under the mallet-stroke, while the heavy foil is stiff enough to hold all in place as it is condensed.

As with incisors, the palato-approximal cavity is more difficult in a bicuspid than the labio-approximal. I may therefore better choose the latter for an illustration to show in what its management differs from that advocated in the incisor region. Fig. 131 shows a bicuspid having such a cavity. There might be circumstances under which I should use a screw here, but when we remember that this position is inaccessible, and that a screw renders any case more trying, it is plain that only extreme necessity would urge its adoption. The main reliance here will be to resort once more to labio- and palato-gingival extensions, as shown at *a* and *b*. I should not make much of a groove under the labial wall, but, passing from the palato-gingival extension, I should make a slight groove along the palatal border, deepening it as I approached the crown, till at *c* it became a distinct concavity. Whether caries has involved the crown or not, the sulcus must be cut out across to its opposite extremity, and at that point a deep retaining-pit should be made as advocated and illustrated by Fig. 28, shown also less distinctly in the present figure at *d*. In some instances, possibly because of extreme sensitiveness, or perhaps from the poor shape of the tooth, only very slight retentive shape may be attainable along the gingival part of the cavity. This would necessitate a proportionate deepening of the palatal groove and the formation of a more distinct one under the labial wall, care being taken not to so undermine it as to weaken it. I should call attention here to the fact that these directions exactly contradict those given for the same cavities in the incisors. There the groove, when made at all, was along the wall left standing, while in bicuspids I direct that it shall be made along that side where the greatest loss has occurred. The latter principle is the more correct, but it is inapplicable in the incisor, for there, there will not be found sufficient space in which to make a groove, which, if it escape the pulp, will not leave the enamel which it undermines so weak that it adds nothing to the strength of the cavity. In the bicuspid it is different. Even were the palatal part of the approximal surface removed till we reach a plane in line with the center of the pulp, owing to the width of the tooth we could still always cut a groove which would make a strong retaining formation, as seen in

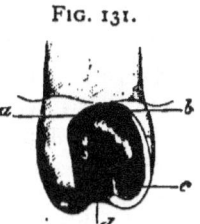

FIG. 131.

Fig. 131 at and between *b* and *c*. In filling these cavities the usual method as described in the previous cases would be followed, save that we should build against the labial wall before the palatal. But in the exceptional cases described, where we get but poor retentive shape along the approximal part of the cavity, we might find it better to begin in the retaining-pit in the crown and build over and into the approximal part, as has been alluded to in the description of the use of heavy foil.

In making contour fillings in bicuspids where the cavity is similar to that in a cuspid, shown in Fig. 126, no special change of plan is to be made, save that where there we had a single undercut in the end of the cusp at *e*, in a bicuspid we may have one in each cusp, and beside resort to the retaining-pit in the sulcus, as in the last figure.

I may now pass to molars, which, though in many respects similar to bicuspids, still present some points of difference. Usually there is so much tooth-substance in all directions that we may almost make the rule that in each given case we shall decide according to the conveniences, and possible dangers of exposing the pulp. In molars more than in any other teeth we should be always cognizant of the fact that *the younger the tooth the larger the pulp*. I remember that once in a first superior molar in the mouth of a miss of twelve, I found an expos-

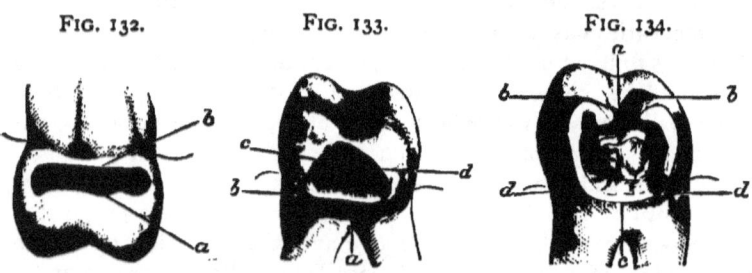

FIG. 132.   FIG. 133.   FIG. 134.

ure of the pulp in a cavity so shallow that without alteration it would not have retained a gutta-percha filling. *Per contra*, late in life we often see molars in which the most extensive undercutting may be attempted with little danger.

On general principles I may say that in approximal cavities in molars I depend chiefly upon two opposing grooves. I prefer that these should lie along the buccal and the palatal borders, but under some circumstances they would be situated otherwise. Such a condition is shown in Fig. 132, in which we observe a narrow cavity extending along the gum-line and partly around the buccal surface. Such cavities often result from the use of ill-fitting clasps, though they may occasionally occur where no clasps or bands have been used. Here we have a condition practically similar to one of the cavities in the bicuspid at Fig. 128. The filling may be retained by grooves along the borders at *a* and *b*, and the gold may be introduced readily with even a small space, being passed in at the buccal aspect. Where a similar cavity encroaches upon the palatal surface, it is much more difficult. If the tooth be on the left upper side of the jaw, I turn the patient's head so that, looking across the roof of the mouth, I may see into the cavity through its palatal extension, and I place the first pellet through this aperture, packing it into the extreme buccal end. From this I build toward and finish at the palatal part. On the right side I follow the same general rule, save that here one must resort more to the mouth-mirror. In the lower jaw all is different. To fill with gold with any hope of success, the widest possible separation is needed, and even then it will often be found difficult, if not impossible, to pack the gold properly should the tooth be long. If the cavity in addition should be on the distal surface, it will be trebly troublesome. Therefore, where I decide to use gold, I most often extend the cavity toward the crown, as shown in Fig. 133, in extreme cases even cutting through to the crown proper. Where such extension is decided upon I do not make much of a groove along the gingival border *a*, but I make a linguo-gingival extension at *b*, at which point I start the filling, and place grooves along the borders at *c* and *d*, of decreasing depth as they approach the crown.

Fig. 134 shows a very unpleasant condition. It is what we term a saucer-shaped approximal cavity, in a molar. Where the dentine is very sensitive, we often find it almost impossible to obtain any anchorages along the gingival border, and, strictly speaking, it is not needed, for the filling can be perfectly retained without disturbing this part of the tooth other than to cleanse it of all decay. Once more I should depend upon lateral grooves along the buccal border, and the palatal, or lingual if it be the lower tooth. But these grooves would be the reverse of those in Fig. 133. There they were deepest nearer the gingival portion. Now I should begin by making an extension into the crown as shown at *a*, with lateral wings or dovetails indicated at *b,b*. From these dovetails I would carry my grooves deep into the tooth-substance, decreasing the depth until they emerged at *d,d*. Between these points *d,d* we have the gingival border *c*, which I have said may be the most sensitive part of the dentine, and we need no undercutting. But where the dovetailing into the crown would leave the lateral walls weak, then we will simply be forced to abandon that plan, making the lateral grooves as deep as consistent with strength, and forming a gingival groove as far as we are enabled.

Fig. 135 shows the loss of the buccal corner as well as the palatal, so that a full contour is required. Where only one corner is absent, the filling may be anchored chiefly into a deep undercut into that which remains standing; but where both are absent, as in the figure, the retaining arrangement is sufficiently unique to excuse a special description. I should make a gingival groove ending in deep dips at each end, similar to the palato- and labio-gingival extensions alluded to in other cases. From these I should extend grooves toward the crown, as shown at *a, a*. It is the formation of these which is unique. I take a rose bur first, and make the groove as deep as determined; then I follow with a wheel bur larger than the diameter of the rose

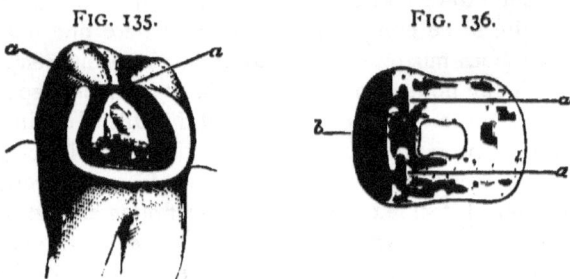

FIG. 135.   FIG. 136.

bur. This wheel, passing from end to end of each groove, forms a lateral undercut in each, which must be made deepest toward the pulp. To better illustrate the idea I introduce Fig. 136, which is a section through the center of a filling placed in such a cavity. The approximal surface is shown at *b*, whilst *a, a* indicates that part of the filling which occupied the grooves. In addition to these retainers, if considered necessary, an extension may be made across the crown and into the opposite sulcus, as already described in discussing bicuspids.

CHAPTER VI.

Special Principles involved in the Preparation of Cavities and the Insertion of Fillings—Cavities in the Masticating Surfaces—Incisors—Treatment of Imperfections—Of Fractures—Of Abrasions—Of Malformations—Cuspids—Bicuspids—Molars—Oxyphosphates in Combination with Gold—Uniting Teeth by Bar and Filling.

I now come to the consideration of cavities in the incisive edges of incisors and cuspids, and the masticating surfaces of bicuspids and molars.

Strictly speaking, a cavity in the incisive edge of an incisor is a rarity, less so in the inferior than in the superior teeth. In making this statement I do not include abraded teeth, but refer to such only as have suffered no other depredation at this part. I have seen a few cases where a distinct cavity, truly carious in character, has presented, seen along the incisive edge only, and not involving either labial or palatal surface. They are found in the form of tiny dark spots, into which the fine point of an explorer readily passes, sometimes to a considerable distance. Such a cavity probably has its inception in an imperfection in the enamel, a tiny pit offering a suitable starting-place for caries. To prepare these, it is only necessary to remove the decay thoroughly, care being observed not to split off the enamel along either surface, thereby producing a fracture which would necessitate an awkward and unsightly filling. To fill, use gold, in tiny pellets, or better still depend upon crystal. When completed it should appear as a small polished gold dot.

I have seen caries along the full length of the edge of a central incisor, appearing as a black line and extending to a considerable depth. When all the decay is removed from such a cavity, no special alteration of shape is needed to achieve retention. The depth, in connection with the narrowness of the fissure, and the roughness of the walls, will prove sufficient. To fill with gold, a convenient method is to select a pellet about the length of the groove, and just thick enough so that it must be slightly compressed as it passes the orifice; this is an exception which marks the difference between general and special principles. Suppose, however, that such a cavity should present, the caries having proceeded only to a slight depth. Then the groove would not be deep enough to retain the filling. Should it be deepened? To do so would be unwise. While we must accept the condition as it is found in the first instance, we must not reproduce it in the second, even though it be an effectual retentive shape. To deepen the cavity would be to render the labial and palatal walls weaker, and make more probable the subsequent splitting off of the one or the other. Such an accident would be especially likely to occur in the mouth of a woman, for despite constant precautions, women will bite off threads with their teeth. The cavity is shown in Fig. 137. It is prepared with a tiny rose bur, and retention depends upon a slight dip into the dentine at each end of the cavity, *a* and *b*. It is evident that this makes the filling after insertion just a trifle longer within, than at the orifice, which is all sufficient. To fill with gold we must proceed differently to the method advocated in the deep groove, where I said we might start with a long pellet. Here it is better to begin at one end, and build across to the opposite pit with small pellets. As soon as the floor is thus

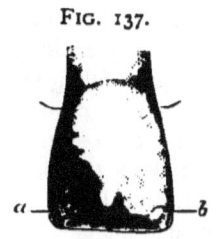

FIG. 137.

covered, and the retaining-points connected, the filling must be completed with heavy gold cut into very narrow strips, about as long as the cavity. The finest-pointed plugger, and the hand-mallet, will give us a filling which will keep its density of surface forever. It is the filling of such cavities as these which show whether the dentist is an artist or merely a mechanic. They are so accessible, and apparently so easy, that I have seen many men at clinics select them for demonstrating rapid work. Yet rapid work here means the use of too large pieces of gold, and imperfect condensation of the gold, with the result that after moderate usage the gold becomes rough. The same system practiced where the entire edge is covered brings the filling back with a ragged edge turned up, as is seen on the ferrule of a cane or umbrella.

Sometimes we find an incisor presenting with a brownish spot which may occur so as to involve the incisive edge. Immediately upon eruption, though the tooth-substance is imperfectly calcified, yet it is protected to some extent by the fact that there is a superficial crust which is more dense than what underlies it. After a few years this crust may break down, and filling become necessary. It is wiser to remove all of this imperfection, as the dentine will be found to be chalky. Where only the labial plate of enamel is involved, the presence of the palatal renders the cavity more simple. I will therefore choose for illustration such a cavity, the preparation of which has required the removal of a part of both surfaces. Fig. 138 shows the cavity in a central incisor prepared for filling. After removing all of the defective dentine, the first step toward retention is to make a rather shallow groove along the whole extent, nearly the full width of the dentine and not encroaching upon the enamel. This direction analyzed is found to indicate that this groove is narrower as it approaches the incisive edge. It is at once seen that already the shape is retentive, since the filling would be wedged in laterally. It would be difficult, however, if not impossible, to successfully fill the cavity so formed, because there is no starting-place. Moreover, should any accident in the future fracture either of the frail corners, the entire filling would be lost, and a difficult operation be entailed for the restoration of the tooth. I have in mind a case where I filled both superior central incisors, having cavities of this nature. Some years later the lady called with the distal corner of one, lost by fracture, and I was able to build on a contour without removing the original filling. For a firm anchorage, therefore, form a dovetail by extensions in the directions indicated by *a, a*, in either of which the filling may be readily started, the arrangement being of such form as will retain the first pellet.

It is opportune at this point to indicate how to proceed should the accident occur to which I have referred, the tooth coming back with one corner missing. It will usually appear as shown in Fig. 139, where we see the retentive shaping as before indicated by the dotted line *a, a*, the filling itself now observed at *b*. The distal corner is absent. A study of this illustration will bring us to the conclusion that the loss of the corner has not impaired the integrity of the filling. Of course had the fracture occurred higher up, let us say on a line with *a*, the gold would need removal. But the most probable presentment is as indicated. How shall we proceed? Very little is to be done to the tooth itself, the main reliance being upon dovetailing into the gold filling. Fig. 140 gives an idea of the new cavity ready for filling. The sharp angle left by the fracture has been removed so that the completed filling will be more sightly. A small rose bur has been passed to the upper end of the extension at *a*, and from this point a groove formed as indicated by the dotted line *c*. A deep undercut has been made into the gold filling, as shown by the dotted line *d*.

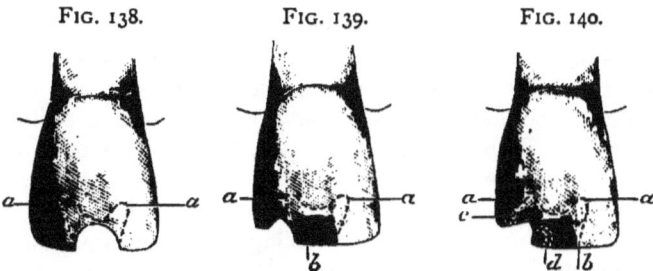

Fig. 138.   Fig. 139.   Fig. 140.

This undercut is made with a bur, and the gold into which the undercutting is carried may be advantageously left rough. It is even possible by thorough drying to produce an approximation of cohesion between the new and the old gold, but this is not essential, as a sufficiently strong mechanical union occurs by careful filling. The first bit of gold may be placed in the extension at *a*, and crystal will be found to serve admirably to begin with.

I will relate an incident which will better convey the importance of properly filling the next cavity which I am about to describe, than any mere words of caution. A child of fourteen whom I saw occasionally, but who was not in my care, fell and splintered off a portion of the cutting-edge and labial face of the enamel of a central incisor, making a cavity which could have been prepared proportionately as shown in Fig. 141. She was taken to her family dentist, and when I next saw the girl a gold filling had been inserted of about the size indicated. I asked permission to look at the tooth, but after examination made no comment, satisfied that any adverse criticism would have been counted professional jealousy. About two years later, having seen little of the girl in the interim, I met her, and my glance at once wandered toward that filling, which I noticed had grown to about the size indicated by the dotted line *c* in Fig. 141. Questioning elicited the information that the cavity had been refilled three times, each operation enlarging the area of gold. The last filling, though a poor one, has remained in place, but the young lady is disfigured for life, or as long as she retains the tooth.

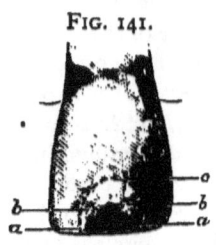

FIG. 141.

When a patient presents with a fracture of this nature, the first care of the dentist should be to examine carefully in order to determine whether the dentine has been denuded at any place. If not, it often occurs that the fracture may be stoned down and polished, even where there may be a considerable concavity, making a much more presentable appearance than after the insertion of gold. The concavity may be made less conspicuous by the judicious beveling of the labial surface near the incisive edge. Even where the dentine has been uncovered, I have sometimes polished it and so left it, whilst in other cases where this seemed inadvisable I have first ground down the tooth, as directed, as far as could be done with safety, and then prepared and filled that part where the dentine had become exposed. Thus it is seen that the aim should always be, first, to avoid filling at all, and second, to make the cavity as limited as possible.

Where the fracture is so deep that a filling is peremptorily demanded, only the finest of burs should be used in forming the cavity. I should recommend that drills be avoided, as tending to shiver and split the enamel. A new rose bur is safest and best. The first step will be to use the corundum, polishing down what need not be included in the cavity proper, and sharpening and perfecting the border lines decided upon. Next, with the rose bur, cut a tiny groove around the semicircle, making it deeper laterally at *a*, *a* in Fig. 141, and deepest at the mesial and distal corners *b*, *b*. This cavity should be first floored over with the tiniest of pellets, and then completed with heavy foil not heavier than No. 30, cut into quite small pieces. Small points are needed, and light taps of a hand-mallet. Such a place as this is a good one in which to avoid a power-mallet of any kind.

In order to make more clear the fact that this plan avoids the danger of disaster such as occurred in the case described previously, I must point out the difference in the arrangement.

When I examined the first filling which had been placed in the child's tooth,.had I criticised it I should have called attention to the following facts, as indicative of future failure: The surface of the gold was pitted, showing that it was not densely packed. Yet other work by the same dentist proved that he knew how to pack gold. The reason of his failure to do so here was seen as soon as the borders of the cavity were examined closely. All around the semicircle the gold could be seen showing through the thin, transparent enamel, whilst the enamel itself showed minute cracks, such as would appear in a bit of china or glass first heated and then dropped into cold water. The probability is, that whilst packing his gold the dentist noted that his borders were breaking down, and deciding that it was the fault of the tooth-structure, he felt obliged to use either a lighter mallet-stroke, or else to depend entirely upon hand-pressure. The truth is, the whole fault lay in the preparation of the cavity, and in showing this I will indicate to the student a cause which is productive of a large proportion of failures along the margins of fillings. What this dentist had neglected to observe, if expressed as an axiom, would read, "*As far as possible, avoid placing gold in contact with the dentinal surface of enamel.*" In the preparation of all cavities, it should be the constant care that *in undercutting, some dentine be left in contact with the enamel*. In the case under consideration the dentist had made a deep groove all around his semicircle, and had entirely removed the dentine from beneath the enamel. Thus as soon as he packed gold into the grooves, especially where he attempted to use the mallet, his gold acted as a wedge to lift the enamel, and, being extremely brittle, it will never endure this. Yet I have myself recommended a groove. But observe that I say "only the finest of burs should be used," and as to the groove itself, my description is, "With a rose bur cut a tiny groove around the semicircle, making it deeper laterally at *a, a* in Fig. 141, and deepest at the mesial and distal corners *b, b*." By "laterally" in the above sentence I mean simply at the sides of the cavity, and do not use the word to describe the formation of the groove except as to depth. I would not make this groove horizontally, but rather obliquely, dipping inward toward the dentine. At the corners *b, b* this dip should be more pronounced, as here we are in safer territory, can reach stronger dentine, and must depend most for the retention of the filling.

In one more important essential had the dentist erred. He contoured the edge of his filling, restoring the incisive edge to its original shape, or made it more square by use of his corundum. When it is remembered that almost every morsel of food passed into the oral cavity is first cut off with the incisors, it is plain that much strain must come upon this weak spot. Where the filling is left square at the incisive edge, it becomes a lever which tends to dislodge the gold. Therefore in finishing such a filling it is always wise to round off the edge thoroughly, even though it be necessary to dress down the full width of the tooth along the labial edge.

Next we come to abrasions. When seen in the earlier stages the edges of the incisors present, worn off on an oblique plane, the dentine being exposed and cupped out so that it appears as a marked depression or concavity. At first this concavity is puzzling. We appreciate the fact that the opposite or occluding tooth is the main factor in the depredation, yet when brought together the offender by no means fits into the grooved dentine. Some have, from this argument, claimed that all these concavities in the masticating surfaces of teeth are results of *erosions*. This is most probably an error. They are *abrasions*, and are accounted for by the fact that the food which is chewed plays a part in the destruction. As long as the end of the tooth is protected by enamel, the abrasion goes on horizontally ; the occluding tooth accurately fits into the abraded surface. *No concavity in the enamel appears. This in itself is sufficient to eliminate the idea of erosion, for erosions invariably form concavities in the enamel before the dentine is exposed, whilst after it is exposed there is no more marked concavity in the dentine than in the enamel, the whole presenting a single continuous cupping.* The contrary is true with abrasion. As soon as the dentine is reached the concavity begins to appear, becoming greater and greater as the dentine is wasted away. *The cause is, that the friction from the food causes a more rapid wearing away of the dentine than of the more resistant enamel.*

I have taken this up at some length, because, while I am not dealing in this work with etiology, it becomes essential to be able to decide between an abraded and an eroded surface, because they require different modes of treatment. An abrasion may safely be filled with gold, while to so treat an erosion is often futile.

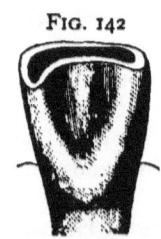

FIG. 142

These groovings in the incisive edges of incisors then are abrasions, and should be filled with gold, or gold and platinum, as early as possible, when filling is demanded. Fig. 142 shows a typical case in a central incisor. The palatal aspect of the tooth is shown, because from this view we may best see the depredation. It is observable that the abrasion is greater at the distal corner. This is frequently the case, due probably to the fact that this corner occludes with two teeth, thus striking against two sharp corners instead of one. Or it may be because, as we are prone to pass the food into one side of the mouth or the other, the distal corner will be given more work than the mesial.

Here we find a tooth needing filling, but having no cavity. It will be requisite for the dentist to be able to thoroughly explain the necessities of the case to his patient, and he should also command his full confidence. The first point to be decided is whether or not the incisors really need to be filled. The only object is to save the teeth from further abrasion, as decay is neither present nor likely to occur. Where the incisors alone are involved, they should be filled. In such a mouth it will be found that in the forward occlusion, as when biting bread, the posterior teeth do not touch. This is a normal condition, and if present, notwithstanding the fact that the incisors are abraded, it will indicate that the anterior teeth have not yet been materially shortened, so that to avoid further loss they themselves must be filled. In a case which has advanced further, when the forward bite is essayed, the posterior teeth will usually be found in contact, though this is not an invariable rule. This will show that the anterior teeth have been considerably shortened. Under these circumstances it may not be necessary to interfere in the incisive region at all, for as the teeth come squarely together regardless of which bite is attempted, it follows that if the posterior teeth be filled in their masticating surfaces further abrasion may be retarded all around the arch. If this be true, it will be unwise to disfigure the front of the mouth by placing gold in the incisive edges of the incisors.

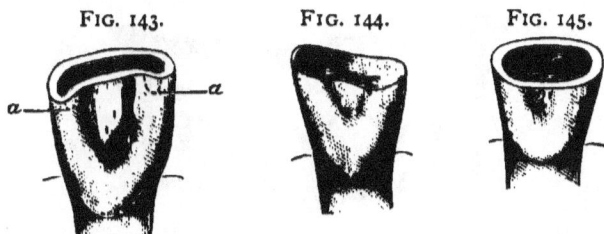

Fig. 143. Fig. 144. Fig. 145.

The preparation of such a cavity as must be made in a tooth abraded as shown in Fig. 143 is as follows. With a corundum stone lightly pass over the labial edge to make it smooth ; then dress down the palatal edge rather freely, designing to restore it with a sufficiency of gold to make a good resistant surface to the action of the opposing teeth. With a rose bur cut a groove, being careful not to reach the enamel in any part. With bur or drill form an extension or pit at each end of the groove as shown by the dotted lines at a, a. In filling, start with pellets of gold foil, and complete with gold and platinum, being extremely careful to thoroughly unite and condense each piece.

When abrasion has been allowed to progress for years, we have that condition which has probably originated the legend among certain folk that their grandfathers had "double teeth all around." A central incisor might present as shown in Fig. 144, the occluding tooth now fitting it like a die, there being little if any concavity. When filling a set of teeth thus destroyed, I deem it unwise to essay the opening of the bite to any extensive degree. The patient is usually old, and will be happier if not asked to adopt a new habit. In the tooth figured, it is essential to completely tip the edge with gold. I should begin by shortening the tooth equal to the amount of gold which I had decided should extend over the edge. This would at the same time bring the tooth into a more symmetrical form, as shown in Fig. 145. The preparation of the cavity is simple, consisting of the formation of a cup, and a groove around to render the shape retentive. There is usually no difficulty from the pulp, which we commonly find either devitalized or calcified. For filling, gold and platinum is preferable to gold, because more resistant.

A somewhat similar form of tooth might occur as a result of fracture, but in that case we should have to consider the presence of the pulp, which now would not be found calcified. We might make some slight undercutting, as shown in Fig. 85, where a failure by another dentist was recorded, but it would be wiser to also rely upon screws, as indicated in Fig. 96, where a lateral incisor was built down.

FIG. 146.

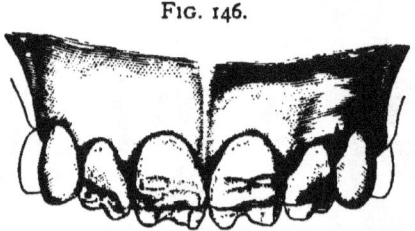

Possibly the most trying cases along the incisive edges are those which appear as seen in Fig. 146, where we have incisors in which the incisive edge is partly absent as a result of fracture, whilst what is left is thin, friable, and malformed. Such teeth have been referred to syphilitic taint, to rickets, and to measles. Whatever the etiology of the condition may be ultimately proven to be, in the meanwhile we have to contend with it as it is. Where neither fracture nor caries has supervened, however unsightly the affected teeth may appear, we *may* decide upon non-interference. When, however, one or more of the teeth are found as in Fig. 146, we are compelled to do something. To attempt a gold filling in the break only would be folly, for the adjacent parts do not offer a reasonable hope of maintaining their integrity. Neither could we place the gold so that it would have a fairly commendable appearance. Most often the serious disfigurement is chiefly confined to the four incisors, extending higher on the centrals than on the laterals. Rarely the incisive ends may be ground off similarly to the method indicated in Figs. 121 and 122. Suppose that we decide against shortening the teeth, and determine upon fill-

ABRASION.

ing, the procedure would be as follows : 'Grind off all the ragged ends of the four incisors, and then prepare each in the form of a groove, in which place screws, three in each central and two in each lateral incisor, as shown in Fig. 147. The fillings should be made with gold and platinum, and the contour restored sufficiently to render the length of the teeth adequate, though we need not always build to original lines. The result is shown in Fig. 148, where we note that the border-line on all four teeth has been made uniform. Slight im-

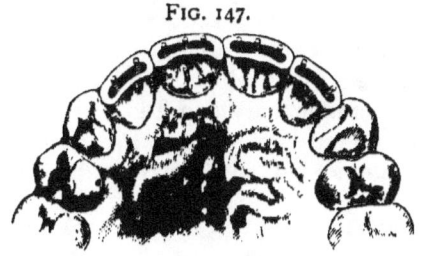

FIG. 147.

perfections still appear in the centrals. These are merely pits which are shallow, and it would be unwise to remove enough of the teeth to eradicate them when they are so high. Some would argue in favor of tipping these teeth with porcelain, while others might advocate cutting off the crowns and replacing them with artificial substitutes. I am not discussing here the relative values of such methods, but must be understood simply as describing how to fill these teeth, when such procedure is deemed best by the operator.

The incisive edge of the cuspid is different from that of incisors,

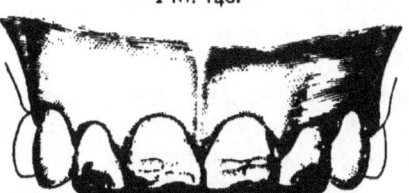

FIG. 148.

and may be considered unique, since it also varies materially from the cusp of the bicuspids, in being more readily cleansed of food because of its being a single instead of a double cusp. Nevertheless, genuine caries will occur with more frequency along the edge of this tooth than in the incisors or the ends of the bicuspid cusps. This will be seen with greater frequency in mouths which show such incisors as I have just been discussing, and the probability is that whilst the cuspids escape the extensive malformation, nevertheless there often occur imperfections in the enamel which, being deep pits, readily serve as an initial point for true caries. In this class of teeth we may find such cavities as is shown in Fig. 149, where we see that the decay has destroyed the edge to one side of the median line of the tooth. This will happen oftener than the cavity shown in Fig. 150, where the actual tip or point is absent. The preparation of either of these cavities is simple, and is indicated by the dotted lines in the illustrations. It consists of removal of all decay, which will usually leave a cup shape favorable for our purpose anyway, but which may be strengthened by slight extensions in opposite directions, rendering the filling, when placed, larger within than at the orifice.

In Fig. 151 we see a cavity which would result were the last one neglected. Its preparation and filling is similar to that of the smaller cavity, except that we must get stronger anchorage, since a more extensive contour is now needed, and, moreover, having approached the

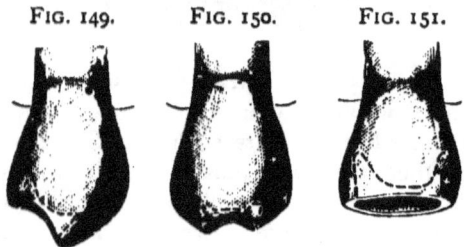

FIG. 149.   FIG. 150.   FIG. 151.

pulp, great care is requisite lest that organ be injured. The main idea after the removal of all decay, is to avoid taking away any of the dentine covering the end of the pulp, which we know is near. Indeed, it is a general fact to be observed in the preparation of all cavities by the methods which I advocate, that *a well-marked hill of dentine appears within the cavity and over the pulp*, which principle has been explained diagrammatically in Figs. 3, 4, 5, and 6. Avoiding, then, the central point of the cuspid cavity, deep extensions are to be made at each side, slanting away from the pulp and lying parallel to the mesial and distal sides of the tooth. Where the cuspid is quite thick through, an advantage will be gained by forming a deep groove connecting these lateral extensions along the palatal part of the cavity. A groove of some depth must be made in any event, for it must be remembered that in mastication the forces exerted will operate to throw the filling outward toward the labial side. This undercut along the palatal side therefore will be a valuable restraint, provided that it is not made at the expense of weakening the wall, a circumstance which should prevent undercutting in any position. All of these cavities in cuspids I prefer to fill with gold. I like its appearance better than the gold and platinum, and in the cuspid, where we have a point and two slanting planes, the force of mastication is much less injurious. A good gold filling in the edge of a cuspid will do faithful service in comparison with gold and platinum in the incisors of the same mouth.

Where the end of a cuspid has been worn off square by abrasion, the one point to be specially noted here is that the tooth is not to be restored to its original shape by reproducing the point. Occasionally it may be necessary to produce a general effect by forming slightly slanting planes rather than to make a perfectly square end, but ordinarily it will be best to simply fill the concavity flush with the highest point of enamel remaining, extending the material over the enamel as a protection. Here, of course, as with the abraded incisors, we must use gold and platinum, for the teeth are now occluding squarely end to end most of the time, if not always.

I pass now to bicuspids, and in doing so I cross that line which by some has been erected as a point beyond which it is permissible to fill teeth with gold or with amalgam, according to the pocket-book of the patient. I have already said that a money consideration is an unscientific standpoint from which to choose a filling-material. Yet I must recognize this line, or at least the territory on one side of it. Nothing would tempt me to fill any incisor or cuspid with amalgam, except in rare cases where the cavity extended under the gum in such a way that my judgment should indicate that gold would fail. Consequently in crossing the line I must admit that I enter the domain of amalgam. Yet even just across the border I do not yet meet amalgam, for I cannot remember to have used it to any extent in the mesial surface of a first bicuspid.

The minute crown cavities found at the extremities of the sulci in bicuspids are extremely important, especially when found in the mouths of young persons. Nothing is easier than to fill these, yet it is an uncommon thing to see them filled properly. What is the reason of this? It will be profitable to discuss it a moment. A well-dressed young miss of fourteen to sixteen, let us say, is brought into the office for an examination. The mirror alone shows large cavities in all four of the sixth-year molars, for the parents, though well bred, stupidly "took them for temporary teeth." Smaller cavities of a similar nature are seen in the twelfth-year molars. Frequently the operator stops his examination at this point, because he has found enough to begin with. He tells the parent that the child has been brought in "in the nick of time, for in a few weeks the pulps would probably have been exposed in one or two teeth." He therefore chooses the worst and makes a start. Perhaps, after all, a pulp is exposed, which means destruction and all the trials and tribulations which follow. Eventually a superficial glance is made at the other teeth, and if an excavator catches in a bicuspid, the parent is admonished to bring her in again in a few months. Now such work on the part of the dentist, though thoroughly conscientious, is not well directed. The bicuspids have been neglected, and are sure to suffer from such a course of treatment. In my opinion, the proper way to examine a young mouth is to begin with a very fine-pointed explorer and examine closely each end of each sulcus, searching for a place, however tiny, where the instrument will penetrate to the dentine. I chose a "well-dressed" miss, for the reason that in such a mouth we would probably find clean teeth, and invisible decay as a consequence. If the explorer discovers a cavity, however small, that tooth must be filled *first*. What about those gaping cavities in the molars? Let them wait, I say. They have been waiting for months, and a day or two more will not make much difference, whilst to begin with them, as I have shown, usually leaves the bicuspids unfilled. The bicuspid with its small cavity, when we consider its immense value in mastication, its ready salvation if immediately cared for, and the extreme difficulties

FIG. 152. FIG. 153.

which it may bring to us later if not saved at the outset, certainly must be counted as worthy of our first attention. Of course there is no need to neglect the molars to their detriment, either. If the work cannot be done coincidently, the decay may be removed from the molars and temporary fillings inserted, whilst we give our first attention to the bicuspids.

Whether the crown of the bicuspid shows one or two cavities, is immaterial. In either event the sulcus must be cut out entirely across from end to end. This I have stated before, and I said that there may be a few exceptions. These will occur more often in the lower jaw, and are confined to that class of teeth in which there is no well-marked sulcus, but simply a pit at each side, the cusps being fused along the median line. In all other cases, extend the cavity to the full length of the sulcus, and, moreover, it must not be cut out with a tiny drill, and filled afterward so that a thread-like streak of gold is all that shows. Fig. 152 is diagrammatic, and is introduced to show the fallacy of this procedure. It gives a section through a bicuspid, and shows the tiny filling lying at the bottom of the sulcus. To so fill a bicuspid gives the operator a chance to exhibit his skill, for it is more difficult to place a solid filling in such a place than where the cavity is enlarged. Early in my career it was my pride to make these thread-like fillings in bicuspids, and in molars as well, following all the windings of the sulci. But it is an error, for it leaves the sulcus practically existent, and in it food will lodge, so that caries will probably recur, the gold dropping out. Fig. 153 shows an enlargement of the cavity, the filling now extending up and filling the sulcus, so that we have left the cusps sufficiently well marked for masticatory uses, whilst we have obliterated the dangerous lodging-place for food and other material. In order to be sure to have sufficient gold in this sulcus, it is well to overbuild, so that after doing so the gold must be cut away to permit normal occlusion. By this means we have as much gold as will be tolerated, and the tooth is safer for it. This rule holds with all crown cavities. The preparation of the cavity to make it retentive is simple. With a reversed-cone bur dip into the cavity at one end, and cut until it emerges at the other. This will form a slight undercut, which may be deepened by following

Fig. 154.   Fig. 155.   Fig. 156.

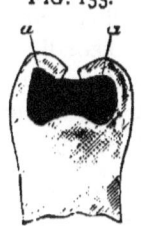
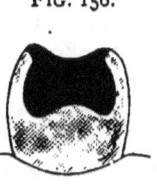

the same course with a wheel bur. Before filling, a fine hatchet excavator should be introduced at each end to make sure that all caries is removed. The depth which will often be reached before this is done will astonish those whose habit has been to neglect bicuspids till the cavities become visible to the eye. For these cavities gold is the only proper material, and it can be placed so rapidly and readily that there is no excuse for the use of amalgam.

Leaving Fig. 153, no special directions are needed relating to simple crown cavities till we come to such as is shown in Fig. 154, a sectional view of which is given in Fig. 155. The entrance to this is along the full extent of the sulcus. If prepared with engine-burs alone, it would be more than probable that much caries would be left unremoved. Generally speaking, it should be remembered that the dentine, as a whole, assumes about the same form as the entire tooth. From this it follows that as the enamel over the cusps forms cones, underlying this enamel in a healthy tooth we would find cones of dentine. Caries finds its way through the weakest part of the enamel, which is along the sulcus, but once it reaches the dentine, further loss of the enamel is rather by a destruction of the dentine, so that the enamel thus undermined fractures under the force of mastication and caves in. Frequently, however, the teeth come to us thoroughly undermined, but with the enamel apparently intact except at the small orifice along the sulcus. Thus the dentine just under the cusps, though decayed, might be left in the cavity unless special care be taken to remove it. A shepherd's-crook excavator might accomplish this, but the result is scarcely desirable, as even where all the carious dentine is successfully taken away the cusps, unsupported by dentine, are likely to crush under mastication later on. I therefore advise the free use of the chisel to thoroughly expose the cavity, and reach strong borders. With a sharp chisel placed at the points a, a, gentle taps of a mallet will remove the overhanging enamel readily and painlessly. Painlessly, because the enamel is taken off in the line of fracture, so that the patient finds no concussion from the mallet-blow. The final arrangement of the cavity is best shown by a diagrammatic illustration, as seen in Fig. 156. This may be considered a section through the tooth, and shows first that the cusps have been cut away so that the filling when inserted covers them sufficiently to afford protection, and secondly we see that in finishing the filling no attempt has been made to exactly reproduce the form of the masticating surface. The filling is merely made to resemble a bicuspid crown in a general way, being left smooth and gently curved, so that it is readily cleansed.

As to selection of filling-material, I should certainly lean toward gold, unless contraindicated by circumstances peculiar to an individual case. To pack the gold I should resort to the method of using oxyphosphate in the bottom of the cavity, pressing in the first pellets whilst the cement was still plastic. This I shall describe more particularly when I come to molars. Where a dentist prefers to use amalgam, it will be seen that no alteration of the cavity will be required. Neither need I give any special directions for placing the amalgam, the cavity being very simple. I should insist on the return of the patient for subsequent polishing.

In Fig. 157 we have a bicuspid from which one cusp has been lost. When this is the palatal cusp it would be as well not to attempt a full restoration, but to be satisfied with an approximate contour, so that the tooth when filled may appear as shown in Fig. 158. This is diagrammatic, giving a section through the filled tooth, and indicating the outer form attained, as well as the relation of the gold to the cavity itself. Where, however, the cusp which is absent is the labial, it will not answer to so proceed, for by not reproducing the full cusp we leave an ill-shaped tooth to attract attention, especially as the occlusion would not be reached. Of course the occlusion would not be reached in the first case either, but being out of sight this would be of no importance. Occasionally we may be able to anchor a filling of this nature so securely that we may be satisfied to risk a full contour. Where it can be done, the procedure is as follows. The removal of all decay would give us a general concavity of somewhat irregular shape.

Begin by forming deep grooves along each approximal wall, as indicated at *a, a,* in Fig. 159. Unite these with a groove around the labial wall, as at *b*. This will give a groove of horse-shoe shape. Make extensions at the labio-approximal angle on each side, as deep as safety will permit. Note that the groove is a horse-shoe, the circle not being completed. To do so would mean to extend the groove along the base of the remaining cusp, which would greatly weaken it, without affording adequate compensation for the risk involved. In fact, except in rare cases, where the standing cusp is found unusually strong, do not attempt any undercutting or grooving at this point at

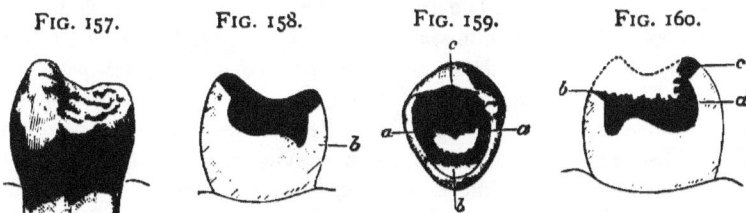

FIG. 157.   FIG. 158.   FIG. 159.   FIG. 160.

all. A general concavity of the whole inner surface of the cusp, as at *c*, will be sufficient to act with the opposing groove, and retain the full contour. Or if not, then one or two gold screws should be employed. A first glance at Fig. 158 might leave the impression that the filling would not remain in place because of the fact that only a single retainer, *b*, is shown. It must be remembered, however, that this, which here is at the palatal side of the tooth, is analogous to the groove at *b* in Fig. 159, where it occurs at the labial. The section does not indicate the lateral grooves, which would occur in such a case as Fig. 158, just as has been described in Fig. 159. As to filling-materials, in such a case as Fig. 159 it seems to me that we must use gold, as the contoured cusp is exposed to view. In Fig. 158, where the whole filling will be unseen, amalgam may be used under some circumstances, as for example in the mouth of one for whom a lengthy operation would be a risk, because of shock. Ordinarily I should use gold in either case. I should begin at the end of the distal groove, and, using pellets, build around till the groove itself was solidly packed. Then I should build across, connecting the two arms of the horse-shoe groove, so covering the floor of the cavity. Keeping my gold with a flat surface, I should build up till on a level with the border where the cusp was absent. Next work toward the remaining cusp, and cover it completely with a thin layer of gold. This is an important point, and that I may be better understood I will resort to a diagram. Fig. 160 shows the tooth in section, filled to a level with the labial wall *b* (supposing that we are dealing with Fig. 159). It is plain that up to this point it has been easy to keep the direction of the plugger and mallet-blow perpendicular to the long axis of the tooth, or in line with the length of the root. By this method the blow receives the greatest resistance, so that the gold is most solidly packed, while the patient reports the least pain. Were we to proceed thus to the end, it would be found that we would be constantly called upon to pay special attention along the standing wall, as for example at the point indicated by *a* in the diagram. We should be compelled to place every piece at that point, first building from there outward, or else risk imperfect packing at this point. In this way we would lose the horizontal plane, the surface of the filling soon becoming oblique. Thus throughout the rest of the operation the mallet-stroke would not be perpendicular to the root. I advise immediately covering the wall as shown in the diagram at *c*. To do this of course necessitates either hand-pressure, or the mallet at an oblique angle. Even the latter, which is preferable, would cause little pain, because the wall would be sufficiently covered with a few pieces of gold. With the cavity filled up to the point shown in the diagram, the completion may be carried on with comparative rapidity, because the tooth-substance being all covered we have no further anxiety in that direction, but may devote our attention exclusively to shaping the contour. From this point use heavy foil, No. 30, or possibly 60, cut in square pieces, and pack from the cusp *b* toward the point *a*. Allow each piece to overlap the cusp *b*, and in this manner the labial portion will grow sufficiently outward, so that surface packing will not be required when the contour is formed. This is a most important point, for we must have the labial surface dense so that we may give it a high polish, yet it is undesirable to be compelled to pack gold on that part of the filling at the end of the operation, because the force of the blow would be at right angles to the tooth, which by this time may have become so tender that this would be exceedingly painful. It can be readily avoided if the caution to build over the edge with every piece be heeded.

There may be some who will argue that I am in error when I state that no special retaining-point or undercut should be made along the remaining cusp. Yet if so, what are we to do where both cusps are lost, as in Fig. 161? In such a case the same method which I described in connection with Fig. 159 may be followed. Here, however, we do not make the groove a horse-shoe, but complete the circle, for we have no standing wall in danger of being undermined. Fig. 161 shows the cavity prepared for filling, the groove being seen at *a*. In Fig. 159 I advised retaining extensions at the labio-approximal angles. In this case they are to be made at the palato-approximal angles also. Thus we would have a filling which, if removed from the tooth for examination (the tooth having been extracted and cracked open), would appear to have four legs to stand upon, which legs would flare outwardly to a slight extent. Fig. 162 gives a section through the tooth and filling, where the retentive arrangement is readily seen. Where the depredation extends beyond this, so that good retaining-grooves and extensions could not be made without endangering the pulp, it would be better to depend upon four screws, which should be thoroughly well anchored when asked to hold so great a contour.

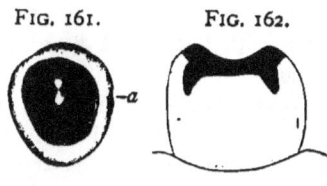

FIG. 161.   FIG. 162.

In molars we find a greater variety of crown cavities. As in bicuspids, there is a class which should receive our immediate attention. I said that with children I usually begin by searching for and filling the cavities in the sulci of bicuspids. This done, I next examine the crowns of the superior sixth-year molars. In the pit at the bottom of the anterior sulcus we are most likely to find at least a small cavity. Call it a cavity, and fill it with gold, if the point of a fine explorer can be made to penetrate so that there will be some difficulty in removing it. When caries has but begun, there is no trouble to fill with gold, for the dam is not a requisite, though always an advantage if it can be used without serious objection. I seldom resort to it with these teeth, because they can be filled so quickly that it is better to depend upon the napkin than to risk too much annoyance to a young patient. Usually a rose bur will sufficiently shape the cavity, cleansing it and making it retentive at the same time. Nevertheless, before filling, it is safer to examine with an excavator, lest some white caries be left. The cavity all ready, fold a large napkin once, and introduce the doubled edge into the mouth and back behind the first molar, where it is held in place with a mouth-mirror, which at the same time reflects light into the cavity. Should there be danger from the saliva flowing from the duct of Steno, a roll of bibulous paper placed between the gums and cheek will suffice to dam it off. Dry the cavity thoroughly with hot air, and select for the first pellet one which will wedge in the cavity. Better still, use crystal. Use a shepherd's-crook plugger if working by the reflection in the mirror, or a slightly-bent bayonet if the patient is tipped back so that the cavity can be easily seen. Depend upon hand-pressure to start the filling, but as soon as the floor is covered and the gold packed is immovable, resort to the mallet. Here the engine-mallet or the electric is most convenient, because one hand is engaged with the mirror. In such small cavities pellets will be more convenient than heavy foil, but the latter should be used at the last. Fill with pellets up to the level with the actual borders of the cavity, and complete with heavy foil, building up so as to lessen the depth of the sulcus.

Next, attention should be given to the posterior sulcus. Here we often find a cavity which will be much more trying, for which reason the rubber is more necessary than in the last case. In this place the decay often extends, though perhaps merely as a line of discoloration, into the palatal groove. When this occurs, or in fact in all cases except where the groove is scarcely defined, the cavity should be extended into it, as is well shown in Fig. 18. It is this extension which will often try the patience of the operator, especially should he be compelled to trust to the mirror for a view of his work. After many and varied experiences, I have decided upon the following method as being the most feasible, and applicable to the greatest number of cases: After fully extending the cavity to the extreme end of the palatal groove, with a sharp spear-drill make a deep pit at this point obliquely, being careful, however, not to wound the pulp. Next enlarge this pit with a rose bur, after which bring the bur forward toward the crown cavity proper, forming an undercut along the sides of the palatal extension. Now exchange for a wheel bur, and pass it the full length of the crown cavity, forming an undercut along each side and at the buccal end. A point of interest is worthy of note here. The caries which decided the operator to fill this cavity most probably was noticed at this buccal end of the sulcus, and is usually found burrowing toward the pulp rather than toward the distal aspect of the tooth. Thus it is essential, after using these engine instruments, to explore for caries at this buccal end of the cavity. In the majority of instances it will be necessary to remove decay, and pain, if experienced at all, will be felt at this time. Thus it is a valuable axiom, that *in all distal or inaccessible crown cavities the mesial side of the cavity should invariably be explored with a shepherd's-crook ..and-excavator*, or other shape suitable for reaching this unseen part of the tooth.

The cavity prepared, the filling should be with gold, and is to be started with crystal in the pit made at the end of the palatal extension. All of this extension may be partly filled until the gold is built well over into the crown portion. It may then be continued with pellets and the mallet until two-thirds completed, when heavy foil should be used in pieces that are narrow, and long enough to extend from the buccal to the palatal end of the cavity. By thus beginning at the palatal end, passing toward and into the crown, finally returning toward the starting-point, the cavity may be rapidly and easily filled. To begin in the crown, however, and build over into the palatal extension, will often be found most tiresome and perplexing.

I have elsewhere designated these two cavities as an exception to the general rule of opening up sulci from end to end. Where considerable caries is present, and the removal thereof discloses the fact that the two cavities are united below the enamel by a narrow passage, I think it best to open up the two into a single cavity, which would then appear as shown in Fig. 163. Here it is to be observed that the two cavities are connected only by a narrow passage. This should only be done where the enamel at a, a is well supported by strong dentine beneath. Where the dentine is carious or absent, they should be cut away freely till a strong wall is reached. This cavity will ordinarily be easily filled with either gold or amalgam, the former, of course, to be preferred, unless contraindicated by circumstances. No

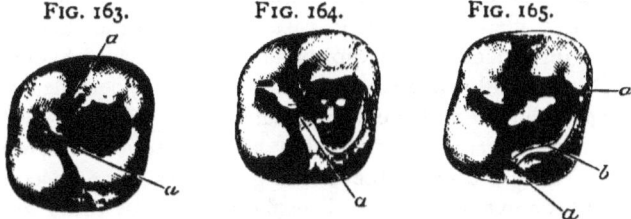

FIG. 163.   FIG. 164.   FIG. 165.

special arrangement is necessary to assure retention, for each cavity usually presenting of general retentive shape, the two when united form the strongest cavity that could be designed.

In connection with a cavity of about the magnitude of that prepared in the last case, we will sometimes find one cusp so undermined by decay, or so imperfectly calcified, that it must be removed in order to reach strong walls. A point of special interest here is the palatal groove. Even supposing that it should be non-carious, it is essential that at least a slight extension of the cavity should be made in that direction. This will give us a cavity shaped as seen in Fig. 164, the antero-palatal cusp being absent. From this it is at once apparent that a point of weakness occurs at a, where we find that all that is left of the cusp is frail. It is therefore obligatory to remove it, so that the cavity ready for filling would appear as in Fig. 165. For the retention of the filling, a deep groove should be formed along the approximal border and around the palatal portion as far as the palatal groove, as indicated at a, a. At the palatal angle b, a depression is to be made as deep as possible without danger to the pulp. Along the base of the standing walls slight undercutting may be resorted to, but care should be taken not to undermine and weaken what is to be the strongest support. This caution especially applies to the antero-buccal cusp, which is in danger of subsequent fracture. The case described being one in which the depredation is supposed to involve the approximal portion only very slightly, it is plain that the dam could be placed save in exceptionally short, or abnormally conical teeth. Therefore I should choose gold as a filling-material. I should start the filling at the palatal angle in the deep dip formed at that point, build backward to the palatal groove, thence over the floor of the cavity and along the walls, and so around and back to my starting-place. All the exposed portion of dentine being thus covered, and all borders perfected, the contour could be rapidly completed by using heavy foil and fairly large instruments with the mechanical mallet. A word as to the contour here. As it is not to show, there is no special object in building the cusp up to the full original height, it being remembered that the more gold the weaker the contour. Were this the buccal cusp it would be more essential to fully restore it, since it would be exposed to view. But in no case need the exact lines of the original be reproduced. An approximate contour will be sufficiently serviceable, and will prove more durable than where deep depressions are made in simulation of the normal sulci.

Where two cusps are absent, if they be the anterior ones, we must observe the same caution at the buccal groove as was advised at the palatal, and remove weak enamel as far as both of these grooves.

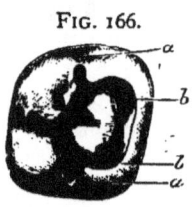

FIG. 166.

Such a cavity prepared for filling is seen in Fig. 166. The retaining groove *a*, *a* now assumes a half-circle, whilst we have two depressions, one at each angle, *b*, *b*, and there is a slight undercut along the standing wall as before. The filling with gold is practically as in the last case. The general arrangement of these cavities will be as described, regardless of which cusps are lost.

In some conditions it may be found advisable to resort to screws, whilst the method shown in Fig. 102 may be adopted in extreme cases. As long as the approximal surface is only slightly encroached upon, however, the fillings can be sufficiently anchored without the screw, or holes cut through the walls, for we could make the groove and retaining depressions of sufficient depth and strength.

Where three cusps are absent, the groove is extended still further, and a third depression at the third angle is to be made. With the disappearance of the fourth cusp we have the groove a complete circle, four retaining depressions being needed, one at each angle, and the tooth is filled and the filling retained exactly as described in connection with bicuspids, and illustrated by Figs. 161 and 162.

Whilst I prefer gold for all four of these crown contours, amalgam may be used with success. The dam should be placed, the cavity being shaped the same as where gold is to be depended upon. The amalgam is then packed thoroughly into all undercuts and retainers, being forced into them with balls of bibulous paper as has been described. Supposing that all four cusps were to be restored, thus choosing the filling of the greatest magnitude, in this class, the amalgam should not be mixed too dry. There should be sufficient plasticity to last during the packing without danger of fracture, or flaking off. The full contour would be restored, the top of the filling appearing convex. The next step is to hasten the setting of the mass by the addition of gold foil, as described in connection with the use of amalgam. The foil is cut or torn into small pieces, preferably as thin as No. 3 foil, and a single piece laid over the amalgam. This is then burnished into the filling with a *smooth warm* burnisher, and becomes incorporated with it. This is continued until it is found that the amalgam is hardening. Then the contour may be perfected by carving gently with a right-angled spatulate burnisher. The burnishing of gold is then continued, smaller ball burnishers carrying the foil into the depressions which have been carved out, and rounding and perfecting the lines. This is kept up till the filling is set enough to be dismissed in safety, at which time it will usually appear to be a gold filling, the final pieces of gold retaining their color. This, however, is lost in the subsequent crystallization, so that when seen at the next sitting the usual amalgam color is presented, save perhaps in spots. A peculiar squeaking sound as the burnisher passes over the surface will admonish the operator that the mass has begun to harden. If the filling after setting be thoroughly polished, it will appear very handsome, and will prove a good and durable piece of work.

I must refer to the large crown cavities, where the cusps are intact, but where the cavity itself is so large that further extension in any direction, either for starting-point or for retention, would so weaken the wall as to make subsequent fracture a probability. It is evident enough that such a cavity will retain the filling once we get it in, but how shall we accomplish this with gold? It is in this class of cases that the oxyphosphate method is found most satisfactory. I promised to explain this in more detail, and will do so here. After the removal of all decay, the dam of course being in place, mix oxyphosphate to a *sticky* consistency, and place it in the bottom of the cavity. Whilst still plastic, press into it two or three large pellets of gold foil loosely rolled, and without condensing let the whole rest till the cement sets hard. We then have such an appearance as is shown by Fig. 167. In this illustration, *a* represents the oxyphosphate, which it is observed has been squeezed out over the borders, *c, c,* whilst *b* shows the loosely packed gold bulging out of the cavity. As soon as the cement is set, whilst waiting for which gold may be prepared for filling the tooth, the next step is to condense the gold and remove the excess of cement. This done, the presentation would be as in Fig. 168, *a* once more showing the phosphate, whilst *b* indicates the gold condensed and cemented to the floor of the cavity. It is to be noted that the unfilled portion of this cavity is still retentive, so that whilst the gold is actually cemented to the tooth, no special dependence is placed upon that for keeping the filling in the cavity. To continue this filling care must be taken to cover the phosphate at the exposed points *c*, *c*, so that no particles may be chipped off to interfere with thorough cohesion. This is best done by carefully laying in pieces of heavy foil, uniting them at the center with the gold already in place, and extending them over the phosphate till the entire floor is presented as a gold surface, when the completion of the filling is comparatively easy. Where amalgam is to be used, the procedure is similar, with a noteworthy exception. The appearance

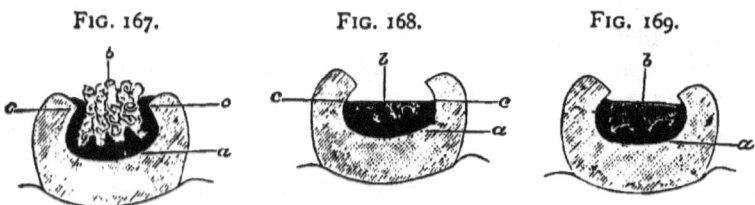

FIG. 167.   FIG. 168.   FIG. 169.

after removal of the excess of phosphate would be as seen in Fig. 169, *a* as before representing the phosphate, which here extends higher up the sides of the walls, whilst *b* is now the amalgam. The completion would leave but a small portion of the amalgam in actual contact with the dentine. Amalgam fillings thus inserted do not discolor the teeth.

In the preparation of a crown cavity in a molar, especially in the lower jaw, it will occasionally occur that in following the direction taken by the caries, we discover that it has eaten its way through to the approximal surface. I have elsewhere said that where a cavity occurs in the crown, and another in the approximal surface of a molar, they should be opened up and filled as one cavity. I did not mean this, however, to include all cases. My language is, "Where a crown cavity in a bicuspid or a molar is but slightly separated from an approximal cavity, the two should be united and filled as one." The word slightly here indicates where I should make an exception to this rule. Where the caries has originated in the crown, and tunneled through to the approximal surface, we will most frequently find occasion for treating differently. The management would depend upon the size of the approximal orifice. If the opening at that point were small, it will be best not to connect the two cavities further than by the tunnel made by the decay.

A condition of this kind is seen in Fig. 170, Fig. 171 showing posterior view of same, and is from a case in practice. A lady presented with a leaky gold filling in the crown of a first molar. Considerable decay was found underlying the gold, which upon removal disclosed the fact that the cavity had reached the posterior approximal surface, the opening there being small and near the gum. This tooth and the adjacent molar were in close contact, and it was desirable, because of the patient's enfeebled health, to submit her to as little annoyance as possible. The idea of wedging, in order to fill the approximal cavity separately, was abandoned. The procedure was as follows. A narrow strip was cut from thin German silver and pressed between the molars. Next a wooden wedge was driven between this strip of metal and the

FIG. 170.   FIG. 171.   FIG. 172.

adjacent tooth, so that it was forced tightly against the opening of the cavity, completely covering it. Amalgam was then introduced through the crown cavity and packed against the German silver. The crown cavity was filled with gutta-percha, and the patient dismissed till the following day. At her next visit the crown cavity was filled with gold, using the method of employing oxyphosphate for starting. This served the additional purpose of separating the gold from the amalgam already in place.

I will introduce a second case from practice, of a somewhat similar character, which received different treatment, though following the same general principles. A lad of fourteen was in my care for the regulation of his teeth, which necessitated the contraction of the arch. As a retainer, bands of pure gold were accurately fitted over the first molars on each side, and to them was soldered a gold wire which, fitting around the arch, restrained all the teeth. Subsequently he came in for an examination, and I found an old amalgam filling in the crown of one of these banded molars leaking badly. Upon its removal I discovered the same condition as last described, the caries having reached the posterior approximal surface. This allowed the gold band to be seen, as is shown in Fig. 172. This retaining fixture having been permanently in place for over a year, I simply filled the entire cavity with gold, building against the gold band at the point where it was visible through the approximal opening. When the time came for removing the retainer, the wire was cut and the band around this tooth allowed to remain in place permanently, appearing as an open gold crown. This will suggest a method which I have in many cases used with success. That is, where a molar is imperfect, or badly undermined along any surface, communicating with a crown cavity, a band such as in the above case may be made and cemented about the tooth, before the insertion of a filling.

Where a crown cavity emerges at the buccal aspect, leaving considerable dentine bridging the part between, it may be filled with a continuous gold filling, by first closing the buccal opening with oxyphosphate and allowing it to set. Then the crown may be filled, and this done the phosphate may be removed from the buccal part and the filling continued. With amalgam this would not be necessary. I once had a case where the pulp had died, the cavity in the crown being quite extensive. Removal of decay disclosed the fact that caries had penetrated the palatal side of the tooth, forming a long groove below the gum-line. A piece was cut from German silver, which, wrapped around the tooth, allowed an extension to pass between the tooth and the gum at the affected point, and so reached the opening. This was held in place by winding flax thread around the tooth, and the cavity was filled with amalgam from the crown, the material being packed against the German silver. This avoided forcing the amalgam through against the gum, and saved the annoyance of passing burnishers below the gum, which when attempted caused profuse hemorrhage.

This will be an opportune place for considering the union of two or more teeth by a single filling. This may be done either for securing a bridge in position, or where, one tooth being loose, we wish to support it by permanently uniting it with its neighbor. It has been claimed by men of repute that these operations can be successfully accomplished by simply holding the bridge-piece, or the bar, as the case may be, until filled around with gold. I have never been able to do this with any degree of satisfaction, or with results of which I felt proud. By the following method, however, I have found the obstacles removed, and success attainable with a sufficient degree of certainty. Where a bridge-piece is to be retained in place by extensions into cavities cut in teeth, held by gold filling, the dam being on, first set the piece with oxyphosphate, and allow the cement to harden thoroughly. Then *remove all of the phosphate from one cavity* and replace it with gold, the piece meanwhile being firmly held in the desired position by reason of its connection with the other tooth. One end secured by gold filling, the same process may be repeated at the other end.

In uniting teeth, use always gold and platinum, further strengthened by a triangular iridio-platinum bar. Amalgam will not serve for these cases, and even gold alone is not durable enough where there is likely to be much strain. The teeth are first to be bound together by flax threads, as shown in Fig. 49, if the teeth are two molars, or as in Fig. 50 when the anterior teeth are involved. If the teeth are quite loose, it will be necessary to imbed them in oxyphosphate, leaving only that part exposed which is to be filled. Next the bar is to be placed in the continuous cavity cut in the crowns of the adjacent teeth, and cemented to place. When the oxyphosphate has thoroughly hardened, as in the case of the bridge-piece, remove it from one cavity and replace with gold ; the bar being thus secured, the rest of the cement may be taken out and the operation completed.

CHAPTER VII.

SPECIAL PRINCIPLES INVOLVED IN THE PREPARATION OF CAVITIES AND THE INSERTION OF FILLINGS—SENSITIVENESS AT THE TOOTH-NECK—EROSION—GREEN-STAIN—TRUE CARIES—FESTOON CAVITIES—THE LABIAL SURFACE—THE PALATAL—THE LINGUAL—BUCCAL CAVITIES—TEMPORARY FILLINGS—THE FINISHING OF FILLINGS.

I HAVE divided cavities into three general classes: approximal, crown, and surface, the last including all parts not covered by the other two. Strictly speaking, however, all three of these terms were intended to apply to that portion of the tooth which is normally covered by enamel. A cavity existing in the root of a tooth, and not owing its inception to caries beginning at some other point, is a rarity, which, with a single exception, needs no particular mention in this work. This exception will be alluded to later.

There is, however, a cavity which in its earliest stages frequently demands our attention; and this I may as well consider now, before taking up surface cavities proper.

Recession of the gum often results in a line of distinct sensitiveness along the neck of the tooth just above the bulge of enamel at the labial aspect of the anterior teeth.

The patient will report the occurrence of pain on brushing the teeth, and that even touching the part with the finger-nail produces an unpleasant sensation. In the earliest stages an examination may reveal nothing except a corroboration of the facts stated, the tooth itself showing no signs either of deep decay or of superficial softening. A filling seems contraindicated in the absence of a cavity, yet the patient may insist upon having relief.

In order that the condition may be thoroughly appreciated, as well as the remedy which I shall advise, it will be necessary to enter somewhat into the etiology of the condition. I have never seen *enamel* sensitive upon its outer surface. If others have noted such a phenomenon, it is sufficiently anomalous to be unconsidered in this connection. I have never seen *cementum* sensitive. Sensitive *dentine*, however, is common. Thus, dentine being the only hard tissue of the tooth which is responsive painfully, it would seem to follow logically that pain at the neck of a tooth must be due to an exposure of the dentine. Fig. 173 represents a sectional view through a cuspid. It

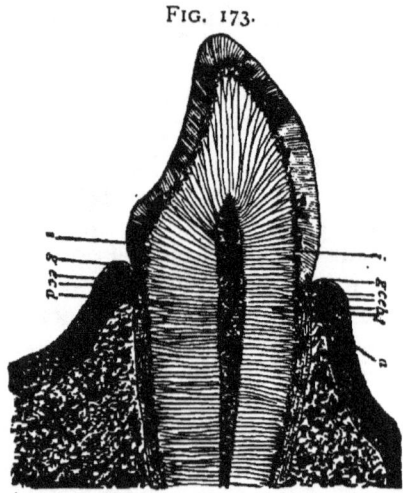

FIG. 173.

is diagrammatical, but is nevertheless sufficiently accurate, since in making the original of the illustration I have produced the various parts by copying drawings which I made from microscopic specimens in the laboratory of Dr. Carl Heitzmann.

The first point of interest to be noted is that normally, at the neck of the tooth, the cementum, *c*, overlaps the enamel, *e*, just beneath the free margin of the gum, *g*. As this gum recedes, possibly as far as the line *a*, the cementum becomes exposed and gradually disappears, thus leaving the dentine exposed at *b*. Very soon thereafter the part may become excessively sensitive. This leads to a consideration of the dentine, *d*, especially at this point. Sensitiveness in dentine has long been attributed to the dentinal fiber. This fiber, though traced as far as the odontoblastic layer of the pulp, has not been posi-

SENSITIVENESS AT THE TOOTH-NECK.

tively seen to anastomose with a nerve-fiber of the pulp itself. That it has the power of transmitting sensation is generally admitted. It has also been universally observed that dentine immediately below the enamel is more sensitive than elsewhere. Exactly at this point we have what is termed the interzonal layer, as at i. The fibers of the dentine bifurcate as they approach this territory and enter it, after which they can be traced as distinct fibers only very sparsely. This might tend to the impression that this interzonal layer should be less sensitive than the main mass of the dentine, but there is considerably more living matter here than in the denser portion, where the tubuli are distinct. Dr. John I. Hart, in an able paper on this subject,* claims that exactly at the neck of the tooth the character of the interzonal layer changes, and that immediately underlying the cementum he finds a granular layer. This I have not myself seen; but the doctor admits that this is very richly endowed with living matter, and so accounts for the sensitiveness.

The mere exposure of this dentine, however, would probably not cause any annoyance to the patient. This leads back to the statement that after the recession of the gum the cementum begins to disappear. I think this is probably due very largely to the tooth-brush, the cementum wasting away by friction. This view is sustained by the fact that though the recession of the gum may occur on the palatal as well as the labial side, the sensitiveness will occur at the labial only. At least I have never observed such a disturbance at the neck on the palatal side of any of the anterior teeth, except when contact with a part of an artificial denture had caused an abrasion or superficial caries.

The destruction of the cementum, then, may be chiefly attributed to the action of the brush, and, therefore, as soon as the dentine is reached it will suffer from the same cause, so that presently the living matter of the interzonal or granular layer becomes exposed. At first it may be responsive only to contact, as when touched by the bristles of the tooth-brush or by the finger-nail. Later it will give pain when subjected to excessive heat or cold, or to the action of acids or sweets.

If taken in this early period, the remedy is as simple as it is effectual. With a clean, smooth burnisher, rapidly revolved in the engine, burnish the affected part, using considerable pressure, and continuing until the patient ceases to shrink. If this be thoroughly done, the operator and patient will be astonished to observe that the part can be freely touched painlessly. How has it been accomplished? Possibly it is that the burnisher drags together the gaping sides of the

---

*See *Dental Cosmos*, p. 723, vol. xxxiii.

minute openings in the dentine, thus closing them and covering up the living matter, which the excessive heat from the friction may also serve to devitalize. Thus a crust is formed over the responsive tissue, so that it is shielded. Fortunately, if the use of a very soft brush be adopted, the sensitiveness will rarely recur within six months or a year, when the remedy may be applied again. Where carious action has actually begun, a superficial softening having supervened, the procedure is slightly different. It is at this time that the tooth will ache after the use of acids or sweets, etc., and an irritation of the living matter has probably occurred. Touch the part with a saturated solution of nitrate of silver, and allow this to remain for a few days. This will discolor the tooth, and, whilst effectual in allaying pain, is a disfigurement. It must therefore be removed with a corundum, or with a polishing-point and pumice. This will at the same time take away the softened dentine from the surface, and the sensitiveness will again be noticed, though in a lesser degree. The burnisher must consequently be resorted to, and persisted in till no softness appears on the surface. This may leave a deep groove, but, especially with women, will be preferable to an unsightly gold filling. Of course, where greater progress has been made, so that a distinct cavity is produced by the removal of softened dentine or decay, a filling is unavoidable.

This sensitiveness at the neck may also occur on the molars; and here we do sometimes find it along the palatal as well as the buccal side. It may also result where attrition has worn down the masticating surfaces till the interzonal layer of dentine becomes exposed. This is a good place for the free use of the nitrate-of-silver treatment. The dam is to be applied, if possible, and the parts touched with the saturated solution, which is allowed to dry. I have treated teeth in this way, sometimes renewing it two or three times at successive sittings, until by the blackening of the surface I was assured that a heavy deposit coated the parts. These teeth not only ceased to be sensitive, but I have noticed that even where carious action had begun, the surface being softened, the caries has been aborted. I have also seen teeth which I had treated in this way five years previously come into my hands again, when, mistaking the discoloration for nicotine stains, I have proceeded to cleanse them. In every such case the patients have returned within a week, complaining that the sensitiveness had returned.

The condition which I have been describing should never be mistaken for erosion,—an error which might be made with the anterior teeth. Yet the distinctions are well marked. With erosions we almost always have the enamel involved. I might say always, for the exceptions are only where the erosion, starting upon the enamel, may spread so as to involve the neck. Again, the erosion will always present with a sharply defined outline, and sensitiveness rarely if ever occurs. This fact is another evidence that the cupping-out of crowns, which are often extremely sensitive, are not erosions. The disturbance at the neck, on the contrary, is sensitive, has no distinct borders, and the enamel is never encroached upon until caries is present.

Surface Cavities.—Cavities upon the surfaces of teeth, other than the approximal, may result from erosion, green-stain, or caries. The last may have as contributory causes, abrasion, malformation, or recession of the gums.

Erosion.—The etiology of erosion is still shrouded in doubt. It is probably a distant symptom of a constitutional disturbance, which has affected the secretions of the mucous glands so that they discharge mucus in which there is an acid present, which has an erosive influence upon enamel and dentine. It is rarely, if ever, seen upon the masticating or approximal surfaces of teeth. It is most common upon the labial surfaces of incisors, cuspids, and bicuspids of the superior set, and cuspids and bicuspids in the inferior. It may occur, however, upon any tooth. Its most distinctive characteristic is that the affected part is hard and highly polished. Erosion may also be said to assume typical forms, of three general varieties. The most remarkable of these is where the disease eats out an acute angle, one line of which is at right angles to the surface of the tooth. This is made plainer by Fig. 174, which gives a profile view of a superior bicuspid, the erosion showing at *a*, whilst in Fig. 175, *a*, we see the same class of erosion in a lower bicuspid. A point which must have some clinical significance, but which I am at a loss to explain, is this : that though the one tooth here is from the upper jaw, and the other from the lower, the erosions are identical. In each case *the line of the eroded angle which is at right angles to the labial surface of the tooth, is at the top, the oblique line extending downward.* In this form of the disease, the pulp-chamber is frequently reached, secondary dentine, however, being deposited, so that the pulp does not become exposed. In a few cases the crown has been cut through so far that it has been lost by fracture. The progress is usually slow.

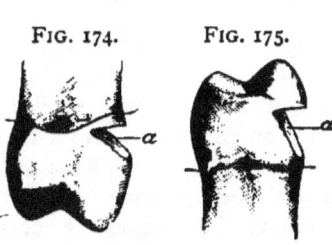

FIG. 174.   FIG. 175.

A second, and to my mind the most dangerous form of the disturbance is an irregular concavity. A patient with a beautiful set of teeth may come in within three months of his last visit, and show small eroded spots on the labial surfaces of several teeth. Within a year they may grow to thrice the size, and in from five to ten years the dentine may be denuded from gum to incisive edge. Unless the constitutional disturbance which is behind this can be controlled, gold fillings serve only a temporary purpose, if they are beneficial at all. Fig. 176 shows a central incisor with a point of erosion on the labial surface. This was filled with gold, and five years later presented as seen in Fig. 177, the filling still in place at *a*, but entirely surrounded by erosion, which continued.

The third form is a uniform groove, which may be deep or shallow in proportion to its width, and may be quite short, or long enough to extend from one approximal surface to the other. They are most commonly seen near the gum. Fig. 178 shows a bicuspid in which is seen the shallow groove near the gum, at *a*, whilst above it, at *b*, is an extreme example of the deep groove, appearing as though cut with a file. It is not difficult to imagine this converted later into the form shown in Fig. 175, by the fusion of the two.

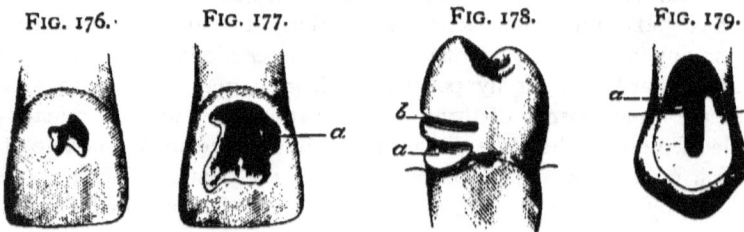

FIG. 176.   FIG. 177.   FIG. 178.   FIG. 179.

Fig. 179 shows a singular case which came under my observation, and is the extreme type of the groove. This is a bicuspid, and had been filled many years previously, the cusps having been restored by gold contour, and the root-canals filled with gold. The subsequent erosion destroyed the labial surface of the tooth, grooving it out until the gold in the canal is fully exposed, showing relatively as indicated at *a* in the illustration. Recession of the gum allowed the erosion to include a part of the root.

The above classification of erosions perhaps cannot be considered as absolute. Erosions vary so much that it might be difficult in some instances to say to which of these types a special case should belong. Nevertheless, what I have described are distinct types, notwithstanding the fact that they may be fused the one into the other.

That they are attributable to different constitutional disturbances is doubtful, as they may all occur in the same mouth at the same time.

Nevertheless, I feel safe in the following statements. The first condition frequently ceases to spread; that is, erosion stops. This is probably due to an alteration in the health of the individual. If this can be definitely known, cavities should be formed and gold fillings inserted. Because of the peculiar shape, it is not difficult to determine this point. The procedure is as follows: Burnish a piece of heavy gold foil, No. 120, into the cavity, allowing the edges to extend slightly over the enamel borders. Then fill with oxyphosphate mixed stiff enough so that in pressing it into place the gold will be more thoroughly adapted to the cavity. When this is set, remove it, and keep it for reference. Fig. 180 makes the method more intelligible, the relation between the phosphate and gold plug, *a*, and the eroded part, being plainly shown. If a year later this plug exactly fits into the cavity of erosion, the disease may be considered cured, and a gold filling is indicated. When it is here admitted that the erosion has not spread in a year, the question may be asked, Why fill at all? The answer is, that though I should wait a year before deciding, I should not neglect to have the patient report during that period. It frequently occurs that after erosion ceases, true caries attacks the point, which is readily understood, since a good lodging-place for *débris* is afforded. Hence the necessity for filling.

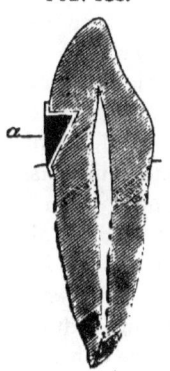

FIG. 180.

In the second class of cases, one must feel that interference by filling can accomplish only one thing. By covering up the eroded surface further erosion may *widen*, but cannot *deepen*, the cavity. If the operator shall decide that this is sufficient gain, it may be accomplished with oxyphosphate as well as with gold, and with less trouble. The final result, where the disease is not checked, will generally be a necessity for artificial crowns.

The groove variety, I think, is more slow in its progress than either of the others. I usually do not fill them until they are deep enough to appear like cavities, when I insert gold, expecting the fillings to be lost in time, explaining the fact to the patient. I have seen such work last so well, however, that I am inclined to believe that it has the advantage of making a slow disease even slower in its destructive action. The formation of all these cavities would be similar to those occurring from caries, which will be described later.

Green-Stain.—Green-stain is a discoloration or stain, greenish in color, found upon the labial, buccal, and occasionally upon the palatal surfaces of teeth. In the mouths of uncleanly persons it would be difficult, perhaps, to indicate a more direct cause than the constant contact of putrefying food. In the mouths of those who would claim to be of cleanly habit, I think I can explain more exactly the presence of this dangerous stain. First, it is to be observed that *the stain is more common upon the teeth of children;* and secondly, that *it will be found more frequently upon the upper than upon the lower teeth.* Both of these are clinical facts which harmonize with my theory. If we pour milk from a glass, we observe that the surface of the glass is left smeared with an oily residuum. Necessarily the same must occur when one drinks milk, allowing it to flow over the enameled surfaces of the teeth. Children drink more milk than their elders, which explains why the green-stain is more common with them ; and as the tumbler or cup is placed to the mouth, the edge rests upon the lower lip, which in turn covers and protects the lower teeth, so that the fluid passes into the oral cavity flowing more freely over the upper teeth than the lower. Thus we should anticipate the stain more often above than below. It is my belief that the fermentation of this residuum from the milk, which adheres to the enamel, produces the green-stain. I have adopted this theory after many years of inquiry, during which I have found more than ninety per cent. of those exhibiting the stain, whether children or adults, admitting that milk formed a large part of their diet. Moreover, I have in a way proved the theory, by having such patients promise to cleanse the teeth thoroughly with brush and hot water after their milk, the result being that the stain has not recurred where the advice has been followed. This I think a sufficient connection between the two as cause and effect, though I do not make the claim that green-stain may not be possible as a consequence of other ferments.

The chief point of interest in connection with green-stain is that, unlike other deposits upon the teeth, this one almost invariably acts destructively. If removed with a rubber disk and pumice-powder, the enamel will be found chalky in color and consistency. Fig. 181 shows a central incisor from which the stain has been removed, the stippling at *a* indicating where the enamel is found decalcified. If taken in time, this will be so superficial that it may be stoned away with a corundum, without penetrating to the dentine, in which case it is advisable to follow that course, and then highly polish the enamel. Later, the underlying surface will be found as in Fig. 182, where we see three small cavities surrounded by an area of decalcification, as indicated by the stippling in the illustration. There is nothing to be done here except to unite the cavities, removing all of the defective enamel, and fill with gold. Fig. 183 is introduced to show how great devastation may be occasioned by green-stain, and is a case from practice, found in the mouth of a miss of ten. This is a central incisor in which a large, irregular cavity appears in the center of the labial face, the boundaries of which, however, do not limit the destruction, the stippling again indicating an area of decalcification. In the mouth in question both centrals were thus injured, and similar though smaller cavities were found in both laterals and both cuspids. The pulps being alive, it was thought best to fill these teeth with gold, notwithstanding the consequent disfigurement. This was an extreme case, yet by thorough attention to the directions as to the care of the teeth, and perhaps more especially because of the abandonment of the excessive use of milk, the teeth have never since been stained, and the fillings are still in good condition, though ten years have passed and the young woman is now herself a mother.

Whilst the lower anterior teeth are more or less protected during the drinking act, this cannot be said of the molar region. The fluid, once in the mouth, flows freely over all the posterior teeth ; and it is along the buccal surfaces of both upper and lower molars that green-stain plays the saddest havoc. It is unfortunately not uncommon to find what at first seems to be a small cavity, the apparent walls of which, however, will be so thoroughly decalcified that they are readily

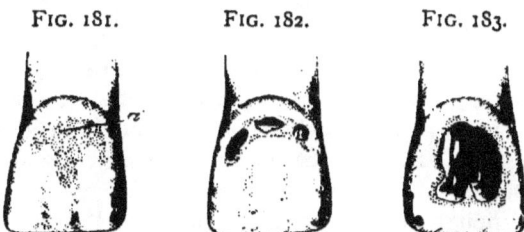

FIG. 181.   FIG. 182.   FIG. 183.

broken down with an excavator. Where this is the case, it is well to first remove the green-stain with a corundum, and thus be enabled to determine what is the real limitation of the destruction. Starting with a small cavity, extension which removes all decalcified structure may result in a cavity which almost, if not quite, encircles the tooth near the gum-margin. I have seen the entire buccal surface so destroyed, with narrow extensions reaching the palatal surface around both approximal sides. These cavities will often prove most exasperating, and their successful filling is doubtful unless one can be assured that future attacks of green-stain will not supervene, which of course is difficult.

True Caries.—Caries may produce surface cavities on any of the teeth, though I have never seen such a cavity in the lingual surface of the lower incisors.* It may follow the depredations begun by erosion, or by green-stain. It may occur in the grooves formed by abrasion after recession of the gum, at the tooth-neck. It may result from malformations, which have made crevices where we should have per-

---

* Since writing the above, a girl of ten presented with a distinct cavity involving only the lingual surface of an inferior lateral incisor.—THE AUTHOR.

fectly smooth surfaces. But most commonly we will find the typical festoon cavity in the anterior teeth, and buccal cavities in molars.

In Fig. 184 is shown the usual festoon cavity as it will be found in the central incisor. The retentive formation has been sufficiently described in Figs. 13, 14, 15, 16. This is a cavity which out of the mouth would be most readily filled, but because of the fact that the upper border is often at, or under, the gum-margin, will be found most trying. Where the gum-border so covers the cavity that it would be impossible to apply the dam successfully, the procedure is to pack a roll of cotton under the free margin as tightly as possible, and then fill the cavity with gutta-percha, so that the cotton will be kept in place. In twenty-four hours the gum will have yielded sufficiently to allow the placing of the dam. Or where the tooth is of favorable shape, fill the cavity with gutta-percha, and slip a ring, cut from rubber tubing, over the tooth and force it under the gum.* This will produce the same result, but should never be attempted without warning the patient that in case she should be prevented from returning at the next appointment, this rubber ligature must be removed, else there would be danger of serious injury to, if not loss of, the tooth. Festoon cavities in the incisors and cuspids of both jaws should be filled with gold wherever possible, and it will be possible in more cases as the dentist forms the habit of making the endeavor to place gold instead of the less reliable, though more manageable, plastics. Where the cavity is excessively sensitive, it may be well at times to fill with gutta-percha, probationally; but within six months, at most, the gold should replace the temporary filling. I have no faith in oxyphosphate or oxychloride in this situation, except for the most temporary purpose,—as, for example, supposing that the regulation of a set of teeth be desirable without delay, and the fixtures would interfere with gold filling. It may be well to fill with oxyphosphate until the regulation has been completed and the opportunity occurs to place gold.

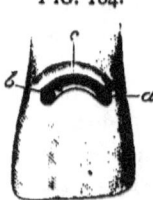

FIG. 184.

A handsome gold filling is less frequently seen at the festoon than in other positions. The cavity itself being easy to fill, this proves how much the proximity of the gum interferes with success.

Taking such a cavity as is shown at Fig. 184, the dam must be placed, and a clamp arranged so that the upper margin is freely exposed to view. Begin by filling the distal retaining pit at *a*, and then fill the opposite one at *b*. Next select a pellet long enough to reach from the gold at one side to that at the other. Fasten at both ends, and then pack it against the upper wall *c*, which

---

* See page 30.

FESTOON CAVITIES.

reference to Fig. 15 will show has no undercut. The absence of an undercut allows the use of a foot-plugger, and the pellet may be packed against and over the upper border. A second and third pellet similarly packed will avoid the danger of slipping of the clamp. Besides, by this method the most important part of the cavity is filled early, before any moisture may have crept in, and the continuance of the filling is made easy. For this, heavy foil should be relied upon exclusively, and fine points with a hand-mallet make the densest filling.

The most difficult festoon cavity is such as is shown at Fig. 185, where we see a cuspid, the root of which has been exposed by the recession of the gum, and subsequent caries has encroached upon it to the gum-margin. Because of the extensive recession at the labial side, with possibly no recession at the palatal, it will be difficult to fit a clamp so that it will not slip. With one or other of the various forms, however, it may be done. The retentive arrangement is different here. It consists of extensions at a, a, as before, but less deep. Grooves are to be made along the borders b, b, growing shallower as they meet at c. The filling is begun in the distal pit, and extended along one groove toward the point c, then down the next groove, and into the opposite pit. Then, starting at c, pellets may be employed to cover the floor, building across from groove to groove until the bottom of the cavity is completely covered. Then resort to heavy foil, and work from the point c toward completion. Use fine points and the hand-mallet, to obtain the densest surface.

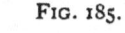

Perhaps I should explain why I prefer the hand-mallet to the engine or electric mallet in these cavities. It is because the best results are attainable with these power-mallets only when a rest for the hand holding them can be had. When working in the crowns of the superior teeth, the hand rests upon the chin. In the lower jaw the power-mallet is less convenient, yet a finger of the left hand may usually be made to serve as a rest or guide to

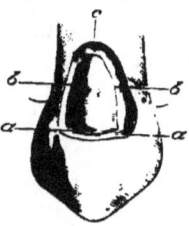

the mallet-point. This is still more difficult in festoon cavities, for which reason I recommend the hand-mallet. The others will make good fillings where the operator has acquired the skill to manage them, but where he finds their use awkward he will probably get a poor result if he persists in using them with the erroneous idea that they make better fillings. Once more I say it is not the instrument, but the man, who accomplishes success.

The preparation of festoon cavities in bicuspids does not vary from that in the cuspids. In the superior jaw they should, in my opinion, always be filled with gold; but in the lower, though gold is preferable, a properly placed and properly polished amalgam filling will do good service.

Where a festoon cavity is connected with an approximal one, the former should be filled first. In these cases, build the gold well up along the gingival border, so that when the clamp is removed to allow access to the approximal cavity, the dam will not slip down. A ligature may then be applied, and, forced above the gold, will easily be tied securely so that all will be kept dry. Where, however, it is decided to fill the approximal cavity with amalgam, the reverse order must be followed. The amalgam must be placed first, and thoroughly polished at the next sitting, when gold may be placed in the festoon, the part which is to be readily seen.

Before leaving festoon cavities, I may mention that on more than one occasion I have seen eight or ten such cavities occur suddenly in the mouth of a person suffering from an acute attack of gastritis. They are also common in connection with a chronic gingivitis, and fillings in such teeth will fail unless the gingivitis is properly treated and cured. Again, there is a species of deposit found in the mouths of persons who may be scrupulously cleanly, which seems to have the power of producing caries. This deposit is most often seen at the labial festoon of the lower anterior teeth. It is whitish, and of a creamy consistency. After properly filling the teeth, the patient's attention should be attracted to this condition, and special brushes recommended, the bristles of which will reach and cleanse these places. Without this precaution the best filling will fail.

Cavities upon the labial surfaces are rare, and mainly due to a malformation, which either produces a crevice, or else poorly calcified enamel. The depredations assume all manner of irregular shapes, but two rules cover them all. Where the cavity is single, prepare it approximately circular, or with the borders of the best curve which will include all of the carious region without unnecessary enlargement. Secondly, where there are two or more small cavities, they are not to be united except they be so near that, filled as a single cavity, the gold will not be more conspicuous than where each is treated separately. (See Figs. 21, 22.) A single illustration will suffice to show the method of management. In Fig. 183 was shown an irregular-shaped cavity surrounded by decalcified enamel. Fig. 186 shows the same prepared for filling. To accomplish this, first use a fine corundum over the surface, removing all of the decalcified territory, which is so superficial that after grinding it away the underlying enamel may be polished. This will leave a clear indication of what the limits of the cavity necessarily must be. The

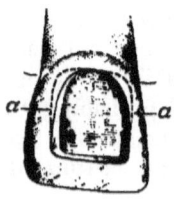

FIG. 186.

FESTOON CAVITIES.   163

next step is to take a sharp excavator, and with a scraping motion remove all of the chalky substance. This will leave the outlines even more clearly defined. Now a sharp rose bur may be used in the engine, and a slight groove made around the whole extent of the cavity. This groove, however, is not at first made in the nature of a retainer or undercut, but is rather to perfect the edges and curve of the border. With the same bur all caries may be cut from the cavity, if any has been left after the use of the excavator. The groove must next be deepened in order to acquire retentive shaping. This is done laterally only, and in a direction sufficiently oblique that the pulp shall not be endangered. The dotted lines at *a, a* show the manner of extending the grooves, and indicate that the deepest points are farthest from the incisive edge. Along the incisive border of the cavity no undercutting is to be attempted; but the original groove must have been deep enough to produce a distinct shoulder here, and avoid anything like a bevel. This also must be true of all edges, which are

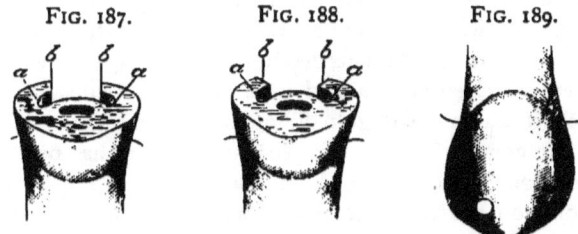

FIG. 187.   FIG. 188.   FIG. 189.

to be at right angles to the labial surface, and not beveled. This can be better explained by showing a sectional view of the right and the wrong way to treat this cavity. Fig. 187 shows a section through a central incisor, in which the cavity has been properly shaped. At *a, a* is seen the deepest part of the retaining groove, and its relation to the pulp and to the orifice of the cavity can be noted. At *b, b* are shown the right-angled edges. In the same kind of cavity, in Fig. 188, the edges, *b, b,* are prepared with a bevel. By this method it is plain from the figure that the retainers, *a, a,* are weaker than in the other, whilst it is also true that the thin edge of gold which will be the result will eventually turn up and break, so that the borders of the filling become defective and leakage follows. Moreover, in polishing, the result pictured in Fig. 47 would easily occur. To fill this cavity with gold, use pellets in one groove and then in the other. This done, build across, uniting them and covering the floor with heavy foil, with which complete the filling. Care should be used not to proceed too rapidly, so that thorough condensation may be had, as well as perfection of border. It is a pleasure, when such a filling is placed, to be able to polish it like a mirror, and have the borders so well made that even under a strong magnifying glass they appear as fine as a hair. This will rarely, if ever, be possible where the beveled edge is depended upon.

Fig. 189 shows a curious cavity which occurred once in my practice, and I introduce it because the method of management will be instructive. This cavity originated in one of those pits which are not uncommon upon cuspids, and which result from malformations. It presented as a small but distinct cavity. On preparing it, to my surprise I noted that, however deep I went, the bottom of the little pit still showed as a black spot. Using a mouth-mirror, I examined the palatal side, and there found a corresponding pit, at a point exactly opposite to that on the labial face. Further exploration disclosed the fact that caries starting in each had met, so that when removed there was a hole completely through, from the labial to the palatal surface. Here was a cavity without any bottom to it. The dam being in place, I stopped up the palatal orifice with oxyphosphate, which I allowed to set hard. I then filled the cavity in the labial face, after which I removed the oxyphosphate and completed the filling from that side, succeeding in obtaining cohesion for my first pellet with the gold packed from the other side, so that when finished there was a solid gold filling from side to side. This idea, conceived in connection with this particular cavity, has been useful in many directions. I have elsewhere alluded to it.

The common cavity at the palatal aspect of the anterior teeth is simple enough, except that great caution is necessary to prevent injury to the pulp. This cavity may be met in either central, lateral, or cuspid, but will most often be found in laterals, occurring in the sulcus. Its preparation is best shown by a sectional view such as Fig. 190, where we see how near any cavity at this point must approach the pulp. I prefer a small rose bur to open the cavity, but any soft decay which may be present must be removed with small spoon excavators. As soon as the limits of the cavity are reached, it will probably be found to be of retentive shape; but if not, extension must be carefully attempted, and must be directed parallel to, or away from, the walls of the pulp-chamber, as seen in the figure. I have seen a palatal cavity neglected till it presented as shown in Fig. 191. If we look at this from a sectional view, as in Fig. 192, we see at once that the pulp is in danger. Nevertheless, this cavity is quite similar to that shown in Figs. 186 and 187 as occurring on the labial surface. As in that case, grooves may be made laterally at *a, a*, escaping the pulp. But, unlike the labial cavity, these grooves can be connected in the palatal cavity, because of the fact that on this side there is a bulge which permits this procedure in safety. The retaining groove, therefore, becomes a horse-shoe, as indicated by the dotted lines *a, a*.

As in the labial cavity, a slight shoulder must be formed toward the incisive portion, in order to avoid a bevel, and so assure a good border to the gold. The same cavity in a cuspid is much simpler. After forming the groove, deep extensions may be made at the gingival angles, owing to the fact that there is a sufficient amount of tooth-substance to make it safe to form strong anchorages. There is another difference between the labial and the palatal cavity, which must be noted. Of two cavities having the same depth, the palatal will reach nearer to the pulp than the labial. This is because the labial surface is convex, whilst the palatal is concave. Consequently in palatal cavities it is often wiser to adopt the oxyphosphate method of starting the gold filling. Where this is done, the gold is pressed into the mass of oxyphosphate toward the incisive edge. When set, the oxyphosphate is removed from the retaining grooves and the gold extended into them, thus securing the filling without special dependence upon the adhesive property of the plastic, which is inserted as an insulator.

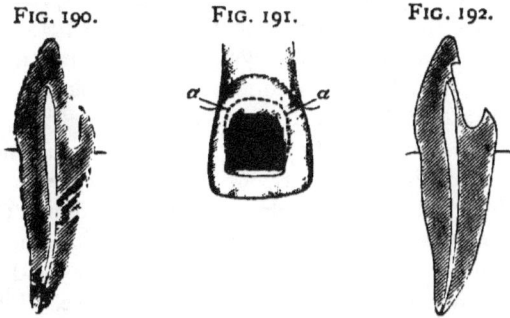

FIG. 190. FIG. 191. FIG. 192.

A lingual cavity in the bicuspids or molars is a rarity, and, when found, usually extends along the gum-border. In nine cases out of ten an ill-fitting clamp has caused the mischief. The lingual surfaces of almost any of the lower teeth may almost be said to be exempt, save where the clamp induces decay. This is perhaps because the tongue and fluids of the mouth keep the parts washed and cleansed. Nevertheless, I have seen long, narrow cavities, partly under the gum, all along the lingual surfaces of lower molars. Their preparation may require that they be packed with cotton for a day, in order that the gum may be forced away. This accomplished, a rose bur which will cleanse the cavity of decay will usually leave it retentive in shape. If not, a slight extension at each end is all that is needed. Here is a place where I might almost say gold should never be employed. To insert a perfect gold filling in such a position, with all the obstacles offered by situation, saliva, and presence of the tongue, would require extraordinary ability, and even then would be accomplished at an expense of time and pain that scarcely excuses the effort. An amalgam filling, on the contrary, may be placed with rapidity and ease, and if properly polished afterward will serve all purposes.

In the superior jaw similar cavities are more common, and at the same time less difficult. I will introduce here a case from practice which offered unusual features. Fig. 193 shows a first molar in which is seen a narrow cavity, *a*, along the palatal side near the gum, whilst a second cavity, *b*, appears at the palato-approximal angle. At first glance one would naturally say that they should be connected and filled as one cavity. This I could not do, for the reason that there was great difficulty in placing the dam. The twelfth-year molar had not sufficiently erupted to retain the clamp, which therefore was necessarily placed over the affected tooth. Again, this tooth was so conical, and both of the cavities were so near the gum, that I found it impossible to place the dam and tie a ligature before placing the clamp. Neither did I succeed in tying a silk around the tooth after placing the clamp. As a consequent result, I found that though I could force the dam back with the clamp so that I could fill either cavity, moisture would leak in through the other. In this dilemma I filled the cavity *b* with gutta-percha temporarily, then placed the dam and filled the cavity *a* with gold, subsequently filling the other. I filled these cavities with gold, for the reason that the teeth were excessively sensitive, and the young man was obliged to wear an obturator, which needed clasps to hold it in place. I felt satisfied that, if filled with amalgam, the gold clasps in contact with amalgam in this special instance would prove mischievous. As I decided to fill the two cavities separately, I could not form any extension in either at the end near the slight separation of dentine at *c*, without undermining that point. I therefore made an extension at the opposite ends in each, and formed slight undercuts along the length of the cavities. In each case the first pellet was packed into the pit at the end, and the gold built forward toward *c*. The patient was tipped back so that I could get direct view of the work looking across the mouth, the tooth being upon the left side. These two fillings were placed three years ago, and though a fixture with clasps has been worn constantly since, no annoyance has been reported.

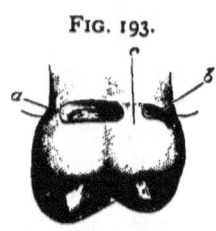

FIG. 193.

Buccal cavities are difficult or simple in proportion as they are large or small, sensitive or otherwise, and near to or distant from the gum. The simplest form is the small, almost circular cavity found in the buccal sulcus, oftener in the lower than in the upper teeth. Their preparation is easy, necessitating the use of a rose bur which will not quite enter the orifice, thus shaping the borders at once as it is pressed through into the dentine. The removal of all carious material, and a slight internal enlargement, will suffice where amalgam is to be used. For gold, I should make slight extensions obliquely in opposite directions; this not so much as a retentive precaution as to facilitate filling, since the cavity so formed will be more readily managed than one which is perfectly regular and round. There is a strong temptation here to wedge in a few large pellets, burnish, and call the tooth filled. The true method is to use small pieces here as elsewhere. I should choose gold for all such cases, except where the youth of the patient might make it advisable not to impose a lengthy operation; then use amalgam, explaining that gold will be inserted later in life.

Where the sulcus is well marked it should be cut out, the cavity being extended toward, but not necessarily into, the crown. This will produce an oblong filling. The retentive formation in such cases would be an extension toward the gingival end of the cavity and lateral, but slight undercutting, effected with a wheel bur. There should be no undercut toward the crown, lest by weakening that

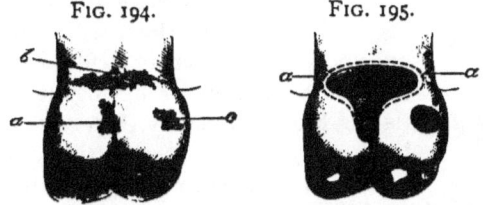

FIG. 194.   FIG. 195.

point fracture result under the forces of mastication. Larger cavities may be met, of all conceivable shapes. There may be two, or even more, distinct cavities in the buccal surface of a single tooth, and in each instance the operator will be called upon to connect them or fill each separately, as his judgment shall dictate. The main fact will be to determine whether the fillings will be better retained separately, or in one cavity. Fig. 194 presents a good study. Three cavities are seen in the buccal surface of a molar. The smaller one, *a*, is in the sulcus, and quite near it is another, *b*, extending along the gum-border. At *c* is a third, which almost encroaches upon the approximal surface, being near the angle. Here it is plain that if in cavity *a* my instruction to make a retaining extension at the gingival end were obeyed, the drill would emerge within the cavity *b*. True, the smaller cavity could be filled, depending alone upon lateral undercutting; but when cavity *b* is similarly prepared it would be found that the separation between the two would be extremely frail. Here, then, it would be best to unite and fill the two as a single cavity. Cavity *c*, however, is distant, and in sufficiently sound territory to permit of filling it separately. Fig. 195 shows the tooth prepared for filling. The dotted lines *a, a*, indicate the retaining extensions depended upon for the larger cavity, whilst in the smaller there is merely a general undercut, slight toward the crown, and deep enough toward the gingiva to allow the first pellet to be wedged securely, if gold is to be used. The choice of gold or amalgam might depend, as before, upon the age of the patient, or it might be decided in accordance with the difficulty met with in placing the dam, and the amount of moisture.

The general arrangement of the larger cavity in Fig. 195 is the one to be depended upon in all ordinary buccal cavities. This is, extensions at the anterior and at the posterior gingival angles, with grooves following the other borders, decreasing in depth as the crown is approached. This applies more particularly to cavities resulting from true caries. Those found under green-stain will be much more perplexing. To illustrate, I will give two examples. The first is seen in Fig. 196, where, the stain having been removed, we find three small cavities along the gum-border, whilst the enamel between and around them is more or less decalcified, as indicated by the stippling. The preparation of this necessitates the free use of a bur, forming a single cavity which will include the entire affected area. This may result in an oblong cavity along the gum-border having nearly parallel borders. More commonly, however, the enamel will be found softened toward the crown along the sulcus, so that, when prepared, the cavity will be approximately shaped like the larger one in Fig. 195. A most disheartening condition is where, whilst extending the cavity in the direction of the decalcified territory, it is found that one or both approximal surfaces have become affected. The rule, however, is not to be relaxed, even though the procedure as directed will lead the dentist entirely around the tooth, thus forming a cavity completely encircling it. This occurred to me once, and I found much difficulty in placing amalgam, until I hit upon a method which led me out of the dilemma. The difficulty was that after packing the amalgam into the posterior approximal part of the cavity, thence into the buccal, and so around into the anterior approximal, as soon as I endeavored to fill the palatal part I found myself dislodging that already packed. Add to this the inroads of moisture, because of the fact that I was obliged to depend upon the napkin, and it is seen that the amalgam, becoming wet, could not properly be forced back into place. I tried beginning at different points, but invariably when I came to completing the circle, dislodgment resulted. Finally I succeeded thus: I fashioned a band of German silver, somewhat wider than the cavity. I began by filling the palatal part of the

FIG. 196.

cavity, packing the material part way into the approximal portions. Next I placed my band in position around the tooth, covering the amalgam already in position, and allowing the two ends to extend forward between the adjacent teeth. Around this band I placed waxed flax thread, the ends lying loose. The condition at this point is shown in Fig. 197, which gives the buccal side of the tooth still unfilled, the relation of the band and flax being seen. The filling was continued from the posterior approximal part around into the buccal, when the band was bent down over it. Then the anterior part was similarly treated. When both ends of the band were thus turned down, the flax was tied, securely holding the band in place, whilst as it was drawn tight it compressed all the amalgam into the cavity uniformly and simultaneously, even forcing out a slight surplus of mercury. This was left in position until the next visit. Since then I have frequently resorted to this method of tying a band around an amalgam filling where similar conditions existed, even though they have been less extensive. It is a good precautionary measure where amalgam is placed in a large but shallow buccal cavity, as well as in many other conditions of anomalous shape, where, after filling, there might be danger of fracture before the mass has hardened.

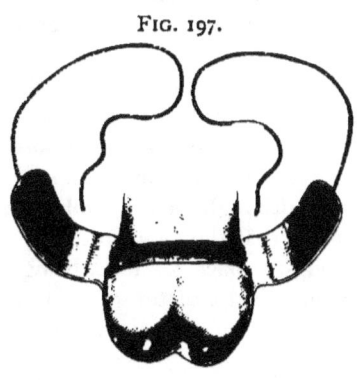

FIG. 197.

The second type of cavities found in connection with green-stain is where the entire surface seems more or less decalcified. Fig. 198 gives an extreme case. Where the depredation is less, so that the crown is not so nearly approached, the borders are simply to be formed into regular curves, and the retaining arrangement will be the same as in Fig. 195, a, a. Where the decalcification has been as great as shown in Fig. 198, it will generally be wiser, if not actually necessary, to extend the cavity into the crown, and then form a groove under all borders. In many instances this will be made simpler by the presence of a filling in the crown, which can then be removed.

FIG. 198.

Only in the smallest buccal cavity, found under green-stain, would I attempt a gold filling. With the larger, and especially where the gingival border passes beneath the gum-margin, I prefer and advise amalgam. An alloy which has a percentage of copper in it has done better service in my practice than any other. No filling, however, will endure, if green-stain supervenes, as the enamel around it will be readily attacked and destroyed, as in the first instance.

There is one other kind of cavity which I promised to describe, though not strictly a surface cavity. In fact, it may scarcely be called a cavity at all, being, if anything, two cavities.

In the mouths of old people, recession of the gum not infrequently progresses to an extent which would seem to menace the tooth. So much of the root is exposed to view that it is marvelous that the tooth should not be loose. Yet often we find such teeth quite firm, although in the molar region the recession has been so extensive that the bifurcation is plainly in view. As a result of this, a lodging-place for food being afforded, occasionally caries attacks the inner sides of the roots, until the pulps being approached some pain is felt, and the patient comes in for relief. Any endeavor to prepare separate cavities in each root would be unwise, for even if successfully filled the space between the bifurcation would continue to act as a repository for *débris*, so that caries would recur. I treat this space as though it were a cavity, simply removing as much caries as possible, when a naturally retentive shape will result. Where the gum is irritated so that it bleeds easily, I like gutta-percha, and prefer the pink to the white. Where the gum is firm, and the pulps not exposed, or else dead, I fill with amalgam. My method is to cut a piece of clean tin foil of such a shape that it can be placed between the roots, covering the gum, thus forming a floor against which to pack the amalgam. When the filling is in, the projecting end of tin is turned up and burnished into the amalgam. In Fig. 199 we see the space between the buccal roots of a molar, and at *a* the tin foil.

FIG. 199.

TEMPORARY FILLINGS.

Passing from permanent, I may profitably discuss temporary fillings. By temporary filling I do not mean a probational filling. The latter term should imply a filling placed in a tooth where some doubt exists as to the advisability of inserting a permanent filling of metal. It is therefore usually of oxyphosphate or gutta-percha. Whilst in a measure intended to serve a temporary purpose, the fact that the tooth is in a doubtful condition of health renders it imperative that the filling should be placed securely and thoroughly, so that it may remain undisturbed as long as possible, thus affording the tooth ample time for full restoration to such a state of health that it will no longer be doubtful whether the final filling of metal may be inserted.

Temporary fillings, then, are those which cover dressings placed within a cavity for medication ; or those inserted to tide over a few days until more convenient to fill ; to force the gum away from cavity-borders, or for similar strictly temporary purposes.

The man who indifferently packs in a temporary filling, with the idea that "it is only for a day," leaving surfaces rough and edges overlapped, is a sloven. Moreover, he is careless of the comfort and interests of his patient. Nowhere is the axiom truer that if a thing is worth doing it is worth doing well.

The most important temporary fillings are those which cover arsenical dressings. Arsenic which is not thoroughly sealed within a cavity may cause serious damage, the more so as it must be allowed to run its course, which may involve the entire bony socket of the tooth, so that the tooth itself is finally lost.* The first caution, therefore, is to observe that where a cavity presents partly filled by hypertrophied gum-tissue, arsenic should not be applied at the first sitting, unless special reason should make it essential, and no need of such haste would excuse the procedure unless the hemorrhage consequent upon the removal of the hypertrophied tissue could be absolutely controlled. Ordinarily a sharp lancet should be used for removing this excessive growth, and a saturated solution of nitrate of silver very carefully applied to the remainder. A dressing carrying some medicament which will act soothingly upon the aching pulp should then be applied, and this covered with a temporary filling. I may as well say at once that sandarac varnish on cotton is a filthy combination to place in the mouth. In a few cases it will serve better than anything else, but in the great majority of instances it can and should be dispensed with. Its main advantage is when it is desirable to force resistant gum-tissue away from a cavity-border, when usually it will serve the purpose perhaps better than more cleanly fillings.

In the case above described I should use the temporary stopping furnished at the depots, which is a combination of gutta-percha and wax. This is to be had of two colors, pink and white. For teeth which will peremptorily need attention at the next sitting, use the pink, whilst when a point is reached where the tooth may be allowed a few days' rest, use the white. In this way the dentist can tell at a

---

*Where serious results obtain from poisoning the gum with arsenic, the treatment is to dress the part locally with tincture of iron (*Tinctura Ferri Chloridi*), and to administer internally the hydrated oxide of iron (*Ferri Oxidum Hydratum*), or better still, the same preparation with magnesia (*Ferri Oxidum Hydratum cum Magnesia*).

glance from the color of the temporary stopping whether the tooth requires immediate attention, or whether it may be passed whilst others are filled.

In the case being discussed, then, the pink stopping would be used. At the next sitting this would be removed, as well as the cotton under it, and it would be found that the gum would be in such a condition that arsenic could be inserted. I will pause here to state that I am not discussing the advisability of using arsenic, but am simply telling how to use it where the dentist does depend upon it. The arsenic being placed carefully upon the point of exposure of the pulp, the temporary stopping must be made so soft that it can be placed without undue pressure. If it seems doubtful that this can be accomplished, it will sometimes be better to use wax, which can be made much more plastic. There is one point to be emphasized. Where the exposure is in connection with an approximal cavity, care must be observed that the temporary stopping is not crowded below the gum, as this will often cause more pain than that experienced from the arsenic. With warm burnishers this filling should be trimmed to proper shape, and made thoroughly smooth. It should approximately restore contour, and should not be so full that it would interfere with occlusion. Often when the cavity is of poor shape, the burnisher may be made so warm that it slightly melts the stopping, when if passed along the borders all around, it will compel the adherence of the material.

There will occur cases where the cavity is of such a nature that though the dentist desires to use arsenic, he will recognize at once that, if covered with temporary stopping, the dressing will most probably be displaced. The procedure in these cases is to have the parts as dry as possible, with dam or napkin as is most feasible, and after applying the arsenic, cover with a thin oxyphosphate, which, adhering to the cavity, leaves an assurance of safety. The cavity can be shaped properly for retaining the permanent filling, of course, the trouble in the first instance being that excavation whilst the pulp is aching would be painful.

After the removal of a pulp, or where the pulp has been dead for any length of time, so that the cavity is necessarily deep, especially in molars, the temporary filling need not be exclusively of temporary stopping. In large approximal cavities, considerable cotton may be placed over that which carries the medicine, and only the outer part covered with the temporary stopping. This renders subsequent removal less troublesome. Again, where a large opening in the crown is present, in addition to a fair proportion of temporary stopping, it will be well to use gutta-percha for the exposed surface, as that material will better withstand the force of mastication.

Where it is desired to fill a large cavity loosely, and yet seal it up sufficiently to keep it clean and protected from the ingress of food, cotton dipped in chloro-percha will be found much better than cotton and sandarac. When cotton and sandarac is to be used, as for pressing away resistant gum-tissue, the cotton should be barely touched to the varnish. Then open the pellet, and fold it again so that the sandarac is inside. This will make a more cleanly plug, whilst giving the outer cotton a chance to swell by absorbing moisture.

THE FINISHING OF FILLINGS.

The final success or failure of a filling largely depends upon the finishing or polishing. A number of points in connection with the various materials are of special interest.

Gold.—A gold filling, when dismissed, should appear like solid metal, smooth at every point and highly polished. Unless there is good reason for postponement, the finishing should immediately follow the insertion of the filling, before removal of the dam. Exception to the rule, so far as the removal of the dam is concerned, would be where its presence would interfere with the polishing, or where nothing is gained by leaving it in place. The judgment of the operator would decide. Approximal fillings should be built out so full that after removing the excess the exact contour would be restored, whilst the surface of the gold would be sufficiently dense to permit the highest polish. This requisite at once banishes the matrix. Whether in the anterior or in the posterior teeth, superior or inferior, I like sandpaper for this work, and prefer the disk on the engine-mandrel to the strip, though both will be required to meet all cases. I believe the more common practice is to use the disk with the sand side facing away from the engine hand-piece which holds the mandrel. Occasionally such a position will be peremptorily needed. Ordinarily I recommend that it be placed exactly the other way. With the sand side facing the hand-piece, and sufficient practice to acquire dexterity, the operator will find that he can manage the greatest variety of fillings. I can reach and polish without other instrument the following: Anterior and posterior approximal surfaces of incisors, cuspids, bicuspids, and first molars, and the posterior surfaces of second and third molars; labial surfaces, especially including festoon cavities, and often the palatal surfaces. After the disk has been used a little so that it becomes pliable, I can trim a filling, shaping approximal angles, without flattening. Where the tooth is long, and wide at the crown, I can reach the gingival margin of the filling by pressing the disk against the adjacent tooth, which compels it to run as though concaved. In many inaccessible places I compel the disk to accomplish my purpose by holding it against the part with a flat burnisher pressed against its reverse side. In fact, the disk on the engine in my hands is more useful than any other finishing appliance. I may say here that in spite of the undoubted ingenuity which has been displayed in the invention of disk-carriers, the best for all practical purposes is still the original simple screw-mandrel. Instead of using a screwdriver, however, the disk may be placed or removed by having the mandrel in the engine, and holding the disk whilst the engine is started quickly.

After using a medium grade of disk for taking off the surplus mass of gold, follow with one made from the finest pouncing-paper. This will produce a good polish. Nevertheless, a higher luster still should be attained by use of a strip of chamois well chalked. For this the best material is what is known as "whiting."

I do not like files, either between the teeth or on labial surfaces. In the former they are apt to leave flat planes, and in any event they make scratches, which must be removed finally with the sand-paper, so that the disk may as well be chosen at the outset. Occasionally a file may be needed for making sufficient space in which to revolve a disk or pass a strip; still, even in these cases I prefer a saw, which, having no cutting sides, removes only the rough excrescences which prevent the ingress of the sand-paper. The approximal trimmer is a humane and valuable instrument. Where the cervical margin reaches to or passes the gum-margin, the over-build of gold should be removed carefully with the approximal trimmer rather than with a sand-paper strip or disk, either of which will so wound the gum that not only will it remain in a state of irritation for several days, but frequently recession is induced. This instrument is also essential for the finishing of festoon fillings. In these cases it will often be better to remove the dam as soon as the filling is placed, for the reason that the clamp interferes with the work. With care, a fine finish may be achieved without wounding the gum. This is, however, a difficult place to contend with. With the approximal trimmer remove the excess of gold carefully, even though that may mean slowly, perfecting the gingival margin and working from the gum toward the incisive end of the tooth, thus avoiding wounding the gum and consequent hemorrhage. The gold trimmed to approximately proper proportions, resort to the smallest disks, placed on the mandrel as before directed. Placing the edge of the disk near the gum, slight pressure will cause it to bend, so that as it is revolved it finishes the surface of the gold, the edge of the disk if required even passing under the gum-margin without wounding it. A final polish may be produced with the polishing-powders on soft-rubber disks, or in connection with wood points. The rubber cup, Fig. 200, is invaluable.

In the crowns of bicuspids and molars, I like finishing burs and burnishers in the engine when the fillings are small. Where they are of medium or large size I use them first, for the reason that a more thoroughly dense surface is obtained by burnishing. But for a final finish I do not admire the irregular surface left by the burnisher. I therefore depend upon small fine corundums, Hindostan and Arkansas stones, finishing with pumice and chalk, used one after the other on the soft-rubber disk.

Amalgam.—The same general rules of aiming at and obtaining a lustrous polish are to be observed with this much-abused material. The approximal trimmer, however, will not avail, and for this reason the gingival margin should be made as nearly perfect as possible whilst the material is yet plastic. The sand-paper disk will do much along this part of the approximal surface, but will not always reach the extreme gingival margin. Here a fine finishing file having a small end, somewhat similar to the approximal trimmer, will serve well, the final finish being given with a spatula cut from orange-wood, and used in connection with the polishing-powders. In crown and large contour fillings, after using the rubber disk with pumice and then with chalk, nothing makes an amalgam filling so handsome as the use of the small engine brush-wheel, of moderate stiffness, used first with moistened pumice and then with dry chalk. I dismiss this class of fillings resembling a mirror in color and luster, and they keep their handsome appearance a long time, remaining forever smooth, though getting dull in time.

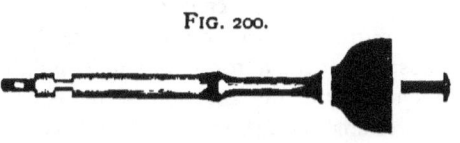

FIG. 200.

Oxyphosphate.—When used as a permanent filling, this material should be allowed to set thoroughly hard before any attempt is made to trim it to shape, and when shaped a final finish may be made with the pouncing-paper disk, followed by the chamois strip without chalk. The dam being still in place, the filling should be thoroughly coated with chloro-percha, which should be allowed to harden by the evaporation of the chloroform before the dam is removed. If this rule is properly observed, this film of chloro-percha will often be found upon the filling two weeks later. Thus the material has been protected from moisture until hardened to a density not otherwise attainable.

Gutta-percha.—This material is trimmed to shape with warm burnishers, and then quickly hardened by applying cold water. After this, if an examination shows that there is an overlap under the gum-margin, dip floss silk in chloroform, and slipping it between the teeth (supposing it to be an approximal cavity), move it back and forth. This will remove the excess without disturbing the filling, as might occur if the use of heated burnishers were essayed again.

CHAPTER VIII.

METHODS OF FILLING THE CANALS OF PULPLESS TEETH—A STUDY OF TOOTH-ROOTS—METHODS OF GAINING ACCESS TO AND PREPARING CANALS—WHEN AND HOW TO FILL ROOT-CANALS.

TEETH are filled that they may be saved ; that is, a tooth which is attacked by caries is in danger of destruction and final loss, and a filling is inserted in the hope of saving it from this result. If a tooth which simply has a cavity in it is in a precarious condition, how much more must this be true where the pulp has died? One might almost say that a tooth which is pulpless is half lost. Its future depends upon the insertion of a proper root-filling ; and where it receives unskillful attention in this direction, only a small chance exists of its long remaining a healthy member of the arch.

The loss of the pulp is not in itself the cause of disaster, for pulpless teeth may remain healthy and useful indefinitely. The trouble is that if the dead pulp be left in, its putrescence becomes a source of excitation which usually results in pericementitis, probably followed by alveolar abscess. The remedy lies in the thorough removal of the pulp, the hygienic cleansing and sterilizing of the canal, and the insertion of a root-filling which will completely replace the pulp, mechanically filling the chamber.

I may, then, at the outset take up a consideration of the obstacles which will hinder the thoroughness of root-filling. There are skillful men, who are also reliable, who will unhesitatingly claim that they fill all roots to, or very nearly to, their apices. This would involve such treatment of the buccal roots of superior molars, and the mesial roots of inferior molars, which are usually admitted to be most difficult. Other men will admit that they are not so successful, but feel assured that they can manage all anterior teeth. That any man practicing dentistry has succeeded in completely filling the canal of every case which he has undertaken, I do not believe. That they may believe that such success has been attained I do not doubt, and therefore I accept such statements as honestly intended, but erroneous from the fact that the gentlemen have not considered teeth

METHODS OF FILLING CANALS OF PULPLESS TEETH. 177

except as they have dealt with them in the mouths of patients, under which circumstances failure to reach the apex may be undistinguishable. A study of roots, out of the mouth, and an attempt to fill the canals, would materially alter the opinions of those who are so certain that they always reach the apex. Nevertheless, I think that these men get nearer to the ideal root-filling than do those who are willing to say quickly, "That is as far as I dare to go," and so fill the canal without having made a conscientious effort to cleanse it. I will now consider the canals of various teeth.

Central Incisors.—The superior central incisor is usually a single-rooted tooth, presenting a fairly straight canal. Nevertheless, it must always be borne in mind that the crown of a tooth is not necessarily a guide to the length, shape, or direction of its root. More true of the posterior regions, this axiom is also true of teeth in the anterior part of the jaw. Fig. 201 represents a central incisor whose crown and root are about proportionate, whilst in Fig. 202 is seen another

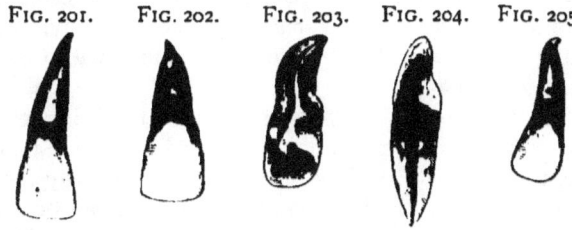

FIG. 201. FIG. 202. FIG. 203. FIG. 204. FIG. 205.

central which has a larger crown and a shorter root. Such short, thick roots are by no means uncommon on central incisors. The point of interest here is, that supposing the dentist is cleansing the canal of a tooth having such a crown as shown in Fig. 202, measurement with a canal instrument might lead him to believe that he had not reached the end of the canal, whereas were he to attempt to go farther he would pass through the apex, forming an opening at the side of the true foramen. I think it may be safely stated that in ninety per cent. of tooth-canals there is a deviation from a straight line just in the foraminal region, so that a drilling instrument would, as I have said, be apt to pass out to one side rather than directly through the foramen. These artificially made openings are almost always mischievous, and the mischief is greater or more uncontrollable in proportion as the drill-hole is nearer the foramen and so more inaccessible than were it nearer the coronal end of the canal. I said that a central presents a fairly straight root and canal; still, there are frequent cases where the root is twisted or curved, an example of which is shown in Fig. 203. In the illustration the palatal aspect of a central is given, and is chosen in preference to the labial, because of the line of approximate bifurcation seen from this view. This makes it possible that the canal within may be divided, the bifurcation of a pulp usually being coincident with that form of root. This figure also shows a distinct curve near the apex, and exhibits the danger that would accompany the free use of any style of drill that has a point allowing it to make forward cutting.

The central in the inferior jaw is usually found with a broad, flattened root, which, viewed from the side, presents a concaved groove extending the full length of the root. This groove is very significant, for it is the lateral wall of the canal, so that it follows that the pulp-canal is the narrowest diameter of the tooth. We must also note that these lateral walls are quite thin. If a canal-reamer were used which had a bur-head larger than this narrow diameter of the tooth, it would follow of necessity that this lateral wall would be punctured, so that it is not alone the *forward* cutting of a canal instrument which offers a danger of opening through the roots of teeth. It would rarely if ever be necessary to use a drill or reamer in the lower incisors, because, as seen in Fig. 204, though the canal is flattened laterally, it is usually wide enough in the other direction to afford ample space for cleansing and subsequent filling.

Lateral Incisors.—The roots of superior lateral incisors almost invariably terminate in a crook at the apex which curves posteriorly. In Fig. 205 is shown a curious example. Judging by the general direction of the crown, the course of the root could not be guessed at all. The root curves toward the median line of the mouth at a considerable angle, yet at the apex the rule above stated is found exemplified, there being a crook which turns posteriorly. It is probable that the root of this tooth was upright when in the alveolus, the crown appearing irregularly curved toward the centrals. Were it not discovered that this curve of the root existed, it is evident that were the root drilled, the instrument might emerge somewhere about half-way between the crown and root-end.

Fig. 206 shows another lateral incisor, where we find a crown not much larger than in the last case, whereas the root is much longer. Again is seen the posterior crook at the extremity. An approximal view of this root would show also an apical curve toward the labial plate of the alveolus.

The lateral incisors of the inferior jaw do not materially differ from the centrals, except that they are slightly larger.

Cuspids.—The cuspid is usually the most readily filled of all pulpless teeth. Ordinarily the canals are proportionately large and moderately straight, and one feels fairly satisfied that at least in this tooth the canal may be filled to the extremity. Yet take a handful of cuspids, and an examination of them out of the mouth will show so many crooked extremities, or ends that assume twists and curves, that a doubt is engendered, and one may well wonder whether even here perfect results are always attained. May it not be that because the canal-explorer reports that considerable length of canal has been reached, the operator decides that he must have come to the apex? May there not still be a crook beyond, which has not been touched by instruments, however fine? Compare Figs. 207 and 208. In both, the

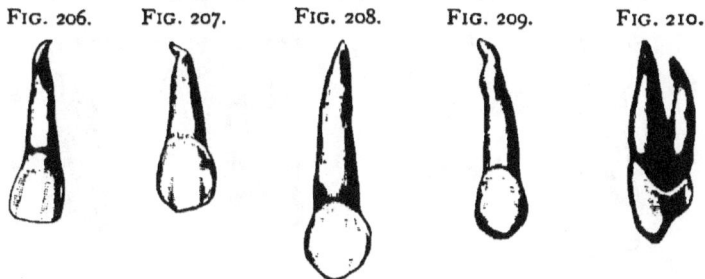

FIG. 206.   FIG. 207.   FIG. 208.   FIG. 209.   FIG. 210.

crowns are about similar in size, yet how different the length of root. In Fig. 207 observe the crook which tips the root end almost at right angles. Look at Fig. 209, with its curved root and crooked end, and, drawing an imaginary line through the central axis of the crown, note where it would emerge through the side of the root were a drill to follow the same course. Fig. 210 shows a double-rooted cuspid, and it would not be difficult to imagine a dentist thoroughly cleansing and filling the labial canal, entirely neglecting the palatal, because of its small size and rarity.

Bicuspids.—The first superior bicuspid brings us many problems in root-filling. Usually the canals are bifurcated, whether the roots are

FIG. 211.   FIG. 212.   FIG. 213.

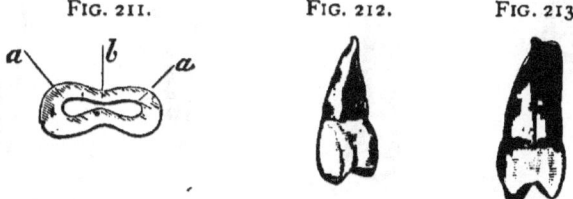

or not. Sometimes the canal will be confluent throughout, being connected by a narrow passage, as shown in the diagrammatic section in Fig. 211. Here a, a indicates the openings to the canals proper, whilst at b is seen a narrow passage connecting the two. It is this passage which is a point of great interest. It is almost always present, at least in the pulp-chamber occupying the crown. It is safe to enlarge it, thus completely connecting the two parts of the canal, until the line of the tooth-neck is reached. Beyond this it becomes necessary to observe the greatest caution in proceeding, in order to determine how far such enlargement may be pursued, whether or not the canals are normally connected throughout, or whether they or the roots are bifurcated. In Fig. 212 is shown a first bicuspid wherein the canals are probably united in this way throughout, as I judge by holding the specimen up to the light so that the canals are indicated, and by the further fact that they emerge at a single foramen. Yet

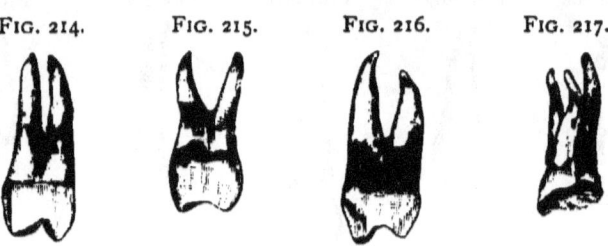

FIG. 214.   FIG. 215.   FIG. 216.   FIG. 217.

observe the curious distortion of the root, and it is plain that enlargement within the canals would be a procedure requiring the utmost care, whilst complete root-filling would be of doubtful possibility.

In Fig. 213 is shown a tooth the roots of which are united, but examination by passage of light discloses the unmistakable fact that the canals are bifurcated from foramina to tooth-neck. Fig. 214 would show similar canals, though here the roots also are completely bifurcated. Either of these cases would be readily cleansed and filled; but Figs. 215 and 216 show conditions where this would not be so simple, as the roots are not only bifurcated, but badly curved.

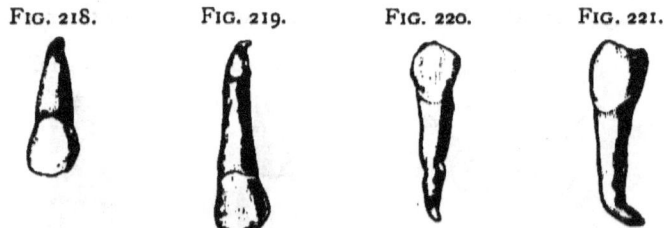

FIG. 218.   FIG. 219.   FIG. 220.   FIG. 221.

The obstacles thus offered are more plainly shown in Fig. 215 than in Fig. 216, for the reason that in the latter the worst curve could only be seen from a different view of the tooth, the roots being bent toward the median line of the jaw. In Fig. 217 is seen a root from which the crown has been lost. Here we have trifurcation which produces three roots, and it is plain that the two smaller would be quite troublesome.

The second bicuspid is usually single-rooted, though even here two canals may be distinctly existent. Figs. 218 and 219 indicate that similar crowns may have roots quite dissimilar as to shape and length. The latter has a tiny but distinct right-angled crook at the foramen.

Inferior bicuspids are often difficult because of the length and attenuation of their roots. I do not mean that this is always so, yet Fig. 220, with its small crown and long, narrow root, is a fair example of a lower bicuspid, while Fig. 221 adds to the difficulties of the situation by possessing an extensive crook.

Molars.—The buccal roots of the superior molars present probably the most difficult problem in the whole range of root-canal filling. To the prominent gentlemen who have repeatedly asserted that they can fill any buccal root, I offer the pair exhibited in Fig. 222 for study and consideration. To these same gentlemen, and to those who are sure that at least they can fill the palatal root, I offer Fig. 223. Supposing for an instant that they succeed in overcoming the obstacles offered by the crook at the end of each of these roots, I would still ask how to fill the canal of the concrescent tooth seen attached to the

FIG. 222.   FIG. 223.   FIG. 224.   FIG. 225.   FIG. 226.

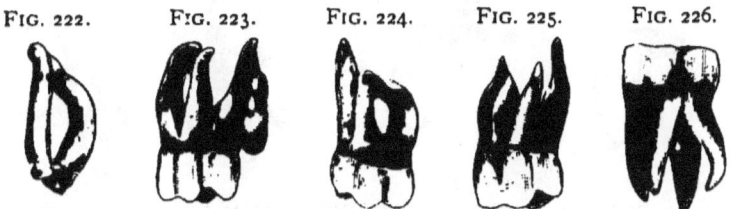

palatal root. Lest it be said that such conditions are quite rare, I introduce Fig. 224, wherein the buccal roots are almost identical to those in Fig. 223.

In Fig. 225, even supposing the dentist succeeded in filling the badly crooked attenuated palatal root, and the equally distorted main buccal roots, he might overlook the little extra root found between the buccal roots.

In the lower jaw it is usually the anterior molar root that is troublesome. Ordinarily we expect single canals, though in the anterior root the canals are often bifurcated. In Fig. 226 is a specimen wherein the posterior roots are completely bifurcated, one offering so bad a crook that it would have been quite difficult to fill it. In the anterior root two distinct canals exist, though the root is single. This tooth and its fellow, which is exactly similar to it, I removed from the mouth of a negro boy. They are sixth-year molars. This tendency to complete bifurcation is more common in the posterior root. I have a number of specimens which show the double root posteriorly and the single root anteriorly, being similar in general appearance to Fig. 226, the extra root in all being at the posterior lingual angle.

The wisdom-teeth (*dentes sapientiæ*) are so commonly misshapen that it is easier to get specimens with distorted roots than with regularly formed ones. They are also so generally condemned to the forceps that root-filling is seldom practiced. Yet occasionally it is best to make the effort to save such teeth, and I may be pardoned for introducing one or two illustrations, that the student may get an idea of what he may have to contend with.

Fig. 227, from the upper jaw, would be a puzzle, whilst Fig. 228, also an upper tooth, would be equally so, with its three roots all curved, the palatal one forming almost a bow. Fig. 229 is a fair mate to it from the lower jaw, whilst Fig. 230 indicates that we cannot always be sure that a wisdom-tooth is short-rooted. The length of the roots in this case would be almost as great an obstacle to thorough cleansing and filling, as would the curve of the preceding specimen.

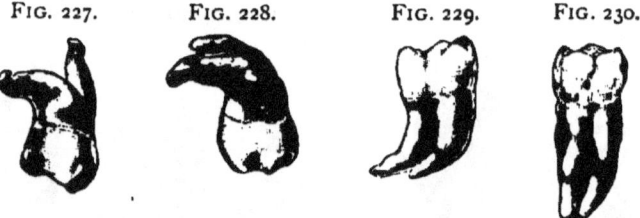

FIG. 227.   FIG. 228.   FIG. 229.   FIG. 230.

Lest it be argued that I have selected extreme cases for illustrating the difficulties to be contended with in filling roots, I must here reply to such a proposition. In the first place, such conditions are not very unusual or difficult to find. That this is true is made plain by the statement that the specimens used here, with a few exceptions, were obtained through the kindness of Dr. Hasbrouck, who allowed me to look through a small lot of teeth extracted in his office. The specimens were selected in about ten minutes. I could have chosen in many instances much more remarkable distortions, but did not think it necessary. I could readily have enlarged the number tenfold, but that also would have been of no advantage. Of wisdom-teeth there was the greatest assortment, possibly because more of these are sacrificed than any other one tooth. The only real rarity among the foregoing illustrations is the two-rooted cuspid. Yet I once removed such a tooth for a young lady, and on the same day extracted a first bicuspid for her mother which had three roots much more marked than those shown in Fig. 217.

Again, I may defend this exhibit by the argument that we cannot learn to anticipate possible difficulties by a study of simple conditions.

It is only by an appreciation of the fact that the roots of teeth are of all manner of shapes, and that the crown is not a sure indication of what we may meet in exploring the roots, that we can hope to exercise that precaution and attain that skill which will make it possible for us to reach that point where we can even fool ourselves into the belief that we are filling all roots to the apices. Yet it is essential that, however great the obstacles may be, we should endeavor to do this; and I will now try to explain the best modes of so doing.

Methods of Gaining Access to and Preparing Root-Canals.

Before a root-canal can be properly filled, it must be thoroughly cleansed and made accessible for the material which is to be used. Admitting that many conditions might occur, as has been indicated by the foregoing illustrations, where it would be impossible or most difficult to fill some canals, it yet is true that many of these can be approximately well cared for where proper methods are employed, and patience and skill are brought to bear upon the obstacles. Conversely, many simple canals are often improperly filled through lack of skill or from laziness. The first object is to attain free access to the canal.

Central Incisors.—The central incisor having ordinarily a straight canal, usually of fairly large size, should offer few obstacles to proper treatment. The cavity of decay must occur either upon one approximal side, upon the palatal, or upon the labial surface. Where it is upon the palatal, the canal is readily entered from that point. When upon the labial, unless the cavity is well extended toward the incisive edge, so that it is not difficult to get directly into the canal, I should make a new opening at the palatal surface. Where the approximal cavity is small, I should do the same; but where large, a simple extension of the palatal border of the cavity should be made until a nerve-canal instrument could be made to enter the canal without bending.

Thus it is seen that I advocate entering the central incisor from the palatal surface. This would also be the case where a pulp had died from traumatic disturbance, and, no cavity being present, a drill should be passed in at the point indicated. The enamel being unbroken, and therefore resistant to the drill, it is well with a small corundum point to grind off the polished surface, after which the drill will cut readily. The drill should be sharp and small, making a narrow opening to the pulp-chamber, which is afterward enlarged with fissure-burs. A cone bur also does this rapidly. Fig. 231 is diagrammatic, and indicates the relation between the usual opening of this character and the pulp-canal. Through such an opening it might be possible to pass a flexible broach and remove the pulp, but it is evident that to attempt to fill the root might result in improper treatment of that part between the opening, *a*, and the end of the canal at *b*. Consequently, my custom is to insert a sharp rose bur through the opening as far as the wall of the canal, and then bring it forward toward *b*, removing the part intervening. Even with the opening thus enlarged, ready access will often be impeded because of the angle at *c*, so that with a bur I effect further enlargement, till the canal, fully opened, appears as seen in Fig. 232. Where these canals are large enough, they need no further reaming than at the aperture; but where they are distorted, attenuated, or partly stopped up because of deposits of secondary dentine along the walls, I use the reamer as far as possible.

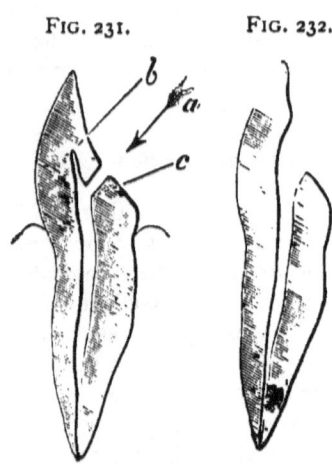

FIG. 231.   FIG. 232.

As the subject of reaming or not reaming canals is one which is much discussed and disputed, and as seemingly equal authorities will most positively adopt one side to the entire exclusion of the other, I may as well here take up this point and give my own views, based upon my experience. The chief objections to using a reamer to enlarge a canal throughout are, first, the danger of making an aperture through the side of the root or near its apex ; secondly, danger of crowding *débris* forward, so that even where it does not pass through the root, carrying possible infection into the territory beyond, it may become packed into the canal itself, limiting further progress, and so after all preventing the proper drilling or filling of the canal ; lastly, the drill may be broken off and remain in the root.

The advocates of reaming all canals say that by enlarging they insure better filling, and that in teeth where pulps have been long dead the canal-walls are probably in a partly softened and certainly septic condition. In these teeth is it not better to *remove* this, than to endeavor to *disinfect* or to *sterilize* it ?

Extremists are not good teachers. The man who reams out every canal to its apex is as unworthy of a following as the other man who can fill all canals without having a reamer in his office. The true method is to have a good assortment of reamers, to know how, when, and where to use them, and to use them with skill and caution that will assure success. I hope that I belong to this last class, and at least I shall describe my methods of using reamers as I go along, indicating where obstacles are to be met and how to avoid them.

I have been asked, "Can you drill around a curve?" The answer is that it can be done, and many times must be done.

Suppose that I am using a reamer, and I feel a resistant spring to my instrument, yet do not strike the end against anything that would indicate that I have reached the region of the foramen. I know at once that the canal has a central curve in it, and that as soon as the head of the reamer reaches this place, the curve of the wall diverting the drill-head, makes a tension upon the flexible shaft so that I get a response which I have described as feeling a resistant spring. This is always a danger signal. Force the instrument a bit farther, and the shaft will snap, breaking off the drill-head. Worse than this, the broken piece will be at such an angle that it cannot be removed. This is made plainer by Diagram 233, where the drill-head is seen at *a*, in a position that makes its removal impossible without very great enlargement of the cavity. Where this happens, it is evident that all the canal beyond must remain unfilled. Consequently, as soon as the springy resistance is noticed the reamer must be removed. If a small-sized Gates-Glidden drill has been used, the largest size may be chosen to replace it. This, having a stouter shaft, may have the power necessary to successfully resist the tension while the drill-head cuts away the bulging wall at *b*. If it was the largest-sized drill that was in use in the first instance, it must be replaced by a fissure-bur of small size, with which the upper bulge at *c* could be removed, after which the large Glidden drill will readily remove the obstacle at *b*. This accomplished by either method, the drill can be advanced as far as *d*, beyond which it cannot be carried. The canal would now present as in Diagram 234. During this work there will be the danger already noted of clogging up by pushing forward *débris*. This may and must be avoided. It results mainly from the desire

FIG. 233.

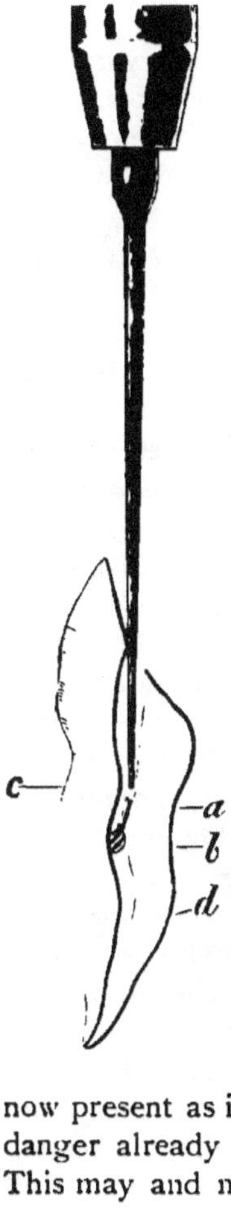

FIG. 234.

to work too rapidly, which, by throwing back a great amount of *débris*, renders it difficult to withdraw the drill. It is compelled to cut its way through, and so leaves within the canal some of the chips, which at the next entrance of the drill are pushed forward and packed into the extremity. The proper method is to avoid this, first by frequently withdrawing the drill, cutting only a little at a time, and secondly by using a clean broach every time the drill is removed, with which the *débris* is easily loosened so that it is withdrawn or may be blown out with the air syringe. The continued use of a drill of the Gates-Glidden type cannot open a hole by forward cutting, since it has a safe end. It could make an aperture by lateral cutting, however, if used where the canal has flattened sides and thin walls. This is most common in lower incisors and in first bicuspids, so that these canals should be most carefully scrutinized before the risk of using a drill is taken. Even where it is deemed safe to use it, by every precaution we should be upon the alert to avoid the disaster. The patient should most certainly give a sign of pain before the opening could be actually formed, for the heat from the friction would be conveyed through the thinning wall, causing a response. Thus it is a safe rule to withdraw a drill upon the slightest evidence that it causes pain.

In Diagram 234 we have the canal partly prepared; but as the curve at the extremity prevented the further use of the drill, what are we to do? What has been gained, since it is after all this very part of the root which we should most certainly fill? The reply is, first, we have lost nothing, for *if it were possible, as some claim, to fill this curved part without enlargement of the rest of the canal, we assuredly can do so now with greater access.* In fact, we can more certainly accomplish it. The use of broaches will indicate the nature of the curve, for as it is withdrawn now, the curve at its extremity, where it followed the bend of the root, will not be bent back, the enlarged canal allowing it to be freely removed. The upper part of the canal being made as clean as possible with broaches, further procedure is with the Evans root-canal drier. This instrument, having a point of soft platinum or silver connected with a copper bulb, may be inserted hot and pliable. The instrument is seen in position in the canal of Fig. 234, and it is observed that the fine silver point is forced almost to the foramen. In many cases it may even pass to the very end. This is accomplished because the heated point destroys impediments in the way of its progress, and being pliable it readily follows curves.

Thus by the judicious use of canal-drills we may enlarge the openings of canals, gaining such access as will render further cleansing more possible, whilst in many instances curves may be entirely eradicated. Consequently, in all cases it is necessary to open a tooth so that a drill may be used, though it will not always be necessary to use one, since such a procedure would sometimes be unwise or even hazardous.

Lateral Incisors.—Lateral incisors are practically the same as centrals. They are, however, smaller, require smaller instruments and greater care, and are more often found with a curved apex. A seemingly straight canal in this tooth, therefore, is to be accepted with greater doubt than where dealing with the central. I have at times withdrawn the Evans root-drier with a sharp crook at its end, showing that it had been forced into a crooked extremity.

Cuspids.—The same rules apply to cuspids as to the incisors. Whilst it is true that these are sometimes short-rooted teeth, ordinarily they have quite long roots, so that the dentist must make a careful examination when his canal-explorers pass but a short distance into the canal of a tooth having a large crown. Where the root is really short, the canal is usually large, the root being thick. Thus it will not be difficult to determine that an abnormal condition is at hand. There is not often much danger of penetrating this root with a Gates-Glidden drill, but the end may sometimes be suddenly attenuated, which will also be the case in the canal, so that the drill may seem to be stopped by the apex of the root, whereas in reality it is simply that the canal has suddenly grown so much smaller that the tip of the drill will not enter it sufficiently to allow the blades behind it an opportunity to cut and so enlarge it. An examination, however, with a fine instrument after the careful removal of *débris* will disclose the fact that the canal continues farther. Perhaps a smaller drill may effect the enlargement, or if not it may be done with a reamer made from a broken Glidden drill. The drill-head being off, the shaft should be stoned on three sides, forming a tapering three-sided reamer. This is a very useful instrument, because it will cut laterally, and is pliable enough to pass curves. But as it will also drill forward, it must be used only with the utmost caution, lest it emerge through the root other than at the foramen. A similar instrument may be made from piano-wire, which, because of its toughness, will not break off in the canal, whilst it will be pliable enough to ream out even tortuous roots to a considerable distance.

Bicuspids.—Generally the bicuspids may be opened for access to the canal either by deepening the crown cavity, or, where it is approximal decay which is present, an extension into the sulcus attains the desired end. In lower teeth, however, this is sometimes troublesome, as will be more fully explained shortly. Where in the upper jaw the cavity is at the neck of the tooth, a new entrance to the canal should be made by entering at the sulcus. Where a tooth, well filled with gold approximally, presents needing to have a canal opened, it will be unwise to remove the filling, as to drill through the crown will serve the purpose adequately and save refilling a difficult cavity.

In the first bicuspid all the peculiarities of form are to be borne constantly in mind, and procedure should be slow and careful. Partial cleansing may disclose the fact that the coronal end of the canal is as represented in Fig. 211. The next point will be to determine just how deep the communicating passage b extends, and therefore how far it will be safe to attempt thorough connection of the two canals, for as two canals they should always be treated at the outset. In a few cases it will be possible to unite them throughout. Most often it will be safe to use a rose bur freely as far as the neck of the tooth. Beyond that explorations should be made with fine broaches, passed up one canal and then forced across into the other, where the passage exists. This will prove a guide, but stiff-shanked burs must be discarded for working farther up in the canals, a slender Glidden drill serving better, because, whilst resistant enough to clear out and enlarge any passage which may exist, it will be found difficult with such a tool to cut through solid material. Where the rose bur or fissure-bur is recklessly used for this place, the inevitable result will be that sooner or later the dentist will make an opening near the beginning of the bifurcation of a double-rooted tooth, or even in one that has only bifurcation of the canals, the roots being coalescent, as in Fig. 213. The palatal canal will not be so difficult to cleanse as the labial, and in this latter, drills are to be used with the utmost caution if employed at all. The position of the patient and the tip of the tooth will generally be such that in forcing the drill into the labial canal it will necessarily bend, and to revolve a slender steel shank under such circumstances is to invite a fracture.

On the other hand, the drill will be more easily used after the coronal end of the canal has been thoroughly enlarged, for the reason that, having more space in which to "play," it will be less likely to bend. Sometimes the cavity in the crown itself may be enlarged, so that the drill will more directly enter the canal. When this can, it should be done. The treatment of curved extremities is the same as with the anterior teeth, the Evans root-drier being the dependence.

The second bicuspid in the superior jaw is usually a single-rooted tooth. Nevertheless, here also the coronal opening will frequently be as in Fig. 211, and once more the narrow connecting passage at b becomes a point for study. This root, though single, is generally broad and flattened, and viewed out of the mouth often has the peculiarity attributed to lower incisors, there being a depressed groove corresponding with the center of the inner canal. Thus it is seen that throughout the canal the walls on either side, corresponding with b in the diagram, are the thinnest part of the tooth. I unite the two broader parts of the canal in a somewhat peculiar way. I use the Glidden drills as though there were two canals, exchanging to larger and larger drills until the two sides are as thoroughly cleansed as possible. This done, I take the smallest of the Glidden set, and, beginning at the coronal extremity of the canal, I pass the drill from one side of the canal through the narrow passage to the other. This is carefully repeated, passing higher and higher up the extent of the canal, until I either clear it throughout or else receive some intimation that it would be unwise to proceed further,—for it must never be forgotten that this tooth also may be bifurcated, at least near the extremity. The least signal of pain from the patient makes it wise to stop work.

The lower bicuspids are often peculiarly difficult because they may be in some abnormal position, the most common and troublesome of which is a tipping inward, so that the labial surface is really partly occluding with the upper teeth. It is plain that in such a pose nothing would be gained by opening through the crown with the hope of using a Glidden drill. Where the cavity is in an approximal surface, however, such extension is often necessary even for the use of a broach, and occasionally the Pettit reamer may be used to advantage. Where the cavity is approximal, but near the gum only, or where it is similarly situated at the labial side, extension must be made along the posterior approximal angle toward, but not necessarily into, the crown. Where it is peremptory to use a drill, as in a case of abscess where the canal is badly narrowed and stopped with secondary dentine, the extension must then follow the center of the labial face, even though it be disfiguring. It is an awkward procedure, but desperate cases demand desperate remedies, and, besides, the labial face of a lower bicuspid is well hidden from view ordinarily.

Molars.—Superior molars having approximal cavities are made accessible by cutting through to the crown. When the cavity occurs elsewhere, but not in the masticating surface, the canals are to be reached not through the original cavity, but through a special opening through the crown made for the purpose. The thorough opening of a pulp-chamber will often require the sacrifice of considerable tooth-substance; but this, though a pity, is unavoidable. No sentimental ideas should tempt the dentist to hesitate to make the opening complete. Where the original cavity is at the posterior approximal surface, some difficulty will be met in cutting through the crown far enough to gain access to the anterior buccal root. The easiest method, and one which will save much distress to the patient, as well as time and labor for the operator, is as follows: With a sharp spear-drill, drill a hole straight to the pulp-chamber through the anterior sulcus. The condition at this point is shown diagrammatically at Fig. 235, which is a section through a molar showing the buccal roots. The approximal cavity and pulp-exposure are seen at a, and the new-drilled hole at b. The next step is to insert a sharp fissure-bur in this hole at b, when, using c as a fulcrum, and slowly tipping the instrument as it cuts, a passage is made through the crown with comparative ease. It must be observed here that we have two advantages by this method. First, by cutting from below upward, the enamel is approached from the dentinal side, which is always easier than to attempt to cut enamel from its outer surface. Second, by using the point c as a fulcrum much less force need be exerted, and that, having a tendency to lift the tooth from its socket, is less painful than the reverse would be. Moreover, there is less danger of having the tool slip when cutting in a hole, than when the effort is made to cut from the approximal surface directly through the crown.

This groove being cut, the presentment from the coronal aspect is as shown in Fig. 236. The next step is to choose a large rose bur, and, passing it in at the posterior opening a, bring it forward, cutting away the dentine at both sides freely, thus undermining the enamel, which latter may then be removed with a chisel. To those who have not essayed this method a close study of the different steps is advised, for by it otherwise difficult and extensive removals of very dense tooth-structure are made moderately easy.

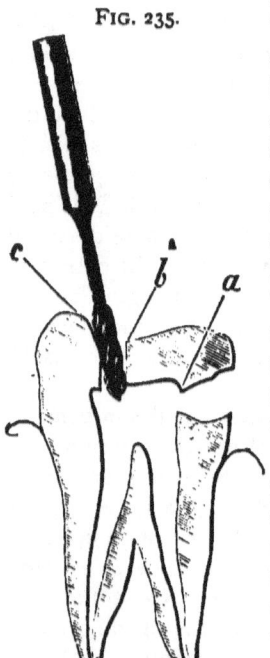

FIG. 235.

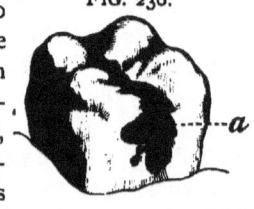

FIG. 236.

Where the cavity is at the anterior approximal surface, the pulp-chamber must be entered with large rose burs, and then, using the anterior adjacent tooth as the fulcrum, the procedure is much the same as before, the enamel being undermined from the dentinal side, and then cut away with chisels.

Practically the same rules hold with the lower molars. Here, as is often the case where the pulp is exposed at the buccal cavity, a drill-hole is made as before in the anterior end of the sulcus until the chamber is entered, after which the entire procedure is as though there were no buccal cavity, except where the latter is so large that it is necessary to unite it with the new crown cavity.

When and How to Fill Root-Canals.

In discussing the actual filling of root-canals, it is essential to consider, at one and the same time, the condition of the tooth and the method of filling its canal ; for despite the fact that many skilled operators have but a single method of treating all roots, I deem it wiser to be guided somewhat by the state of health presented.

Before proceeding, however, I will allude to some of the various materials which have been largely recommended, and comment upon them.

Gold.—Gold at one time was counted the only true material for filling a canal. If the tooth needed gold in the cavity, it also needed this precious metal in the root. The method adopted was to twist a rope of foil stiff enough to allow of its being forced into the canal, and yet soft enough so that it could be condensed thereafter. The method condemns itself, for it is apparent with but a moment's consideration that wherever the foramen was large the rope would be forced through, with the probability of causing future irritation and abscess. This is true of other materials, but in a less degree because of their plasticity. Lest some may claim that I am exaggerating this danger, I may say that I have frequently, in earlier years of practice, removed abscessed roots, finding gold projecting beyond the foramina.

Lead.—Lead has been used considerably, and by some it is claimed that it exerts a therapeutic effect. Just how this is accomplished has never been satisfactorily explained to me, and I doubt its truth. It is used in a single cone, trimmed to shape with the knife, and driven into the root. The same accident of passing through the canal may occur, and again it may be wedged into the upper part of the canal without reaching and filling the foraminal end.

Wood.—Within recent years it has been taught by some that a most excellent way of filling a root-canal is to trim the end of a stick of orange or other wood, and after dipping it into some germicide drive it into the canal, leaving it there. I cannot too strongly condemn this method. Once more we find the material driven through the canal end ; and even where this does not happen, abscesses are common. I have seen a very large number of them. Then, when it is imperative to remove the canal-filling, the operation will prove most trying. The wood splinters, so that it cannot be withdrawn with forceps, tweezers, or pliers, whilst a drill simply tears it to shreds which are still more difficult of removal.

Cotton.—Next to wood, I think this the most despicable thing to leave permanently within a tooth-root. I have heard men tell of removing cotton which had been in teeth for many years, and which had kept the canals sweet, there being a noticeable odor of carbolic acid still present. I always think that the odor is probably due to the fact that the dentist uses that remedy freely, so that he can smell it whenever his own fingers get near to his nostrils. I have removed a great many cotton root-fillings, and have noticed distinct odors in nearly all cases, but they have been far from sweet. They have been of that order which is associated with the dead and the putrescent. In some cases I have unearthed odors which were as vile as anything that had ever assailed my olfactories. I will introduce here a case from the practice of a dental friend which is instructive and suggestive. A physician called upon him, bringing his wife, concerning whose condition the dentist was consulted. The history was that at each menstrual period the woman suffered greatly with neuralgic pains in the uterine region. These increased in severity, and after a time occurred as well in the face. This latter fact, coupled with the time at which the symptoms had first presented, which was directly after having a large amount of dentistry done, suggested to the mind of the husband that possibly the teeth might be the distant cause of all the trouble. My friend made an examination, and finding both central incisors pulpless, hazarded the removal of gold fillings, when he discovered that the roots were filled with cotton. This was removed, and after sterilization they were filled with gutta-percha, and the cavities as before with gold. For six months the neuralgic pains occurred, only in much less severe shocks. As there was some improvement, the husband was encouraged, but decided that there might be other teeth in similar condition, and insisted upon the removal of all the fillings under which there was even a remote possibility of finding cotton. This resulted in such a discovery in three more teeth, and after proper treatment the patient was entirely restored to health. Never fill a root permanently with cotton.

Cements.—The cements, so called, including oxychloride and oxyphosphate, must of course be relied upon in setting crowns, and therefore if they serve in those cases it follows that they must be reliable in any. This, however, must be modified, for where a crown is set, the canal is usually enlarged so considerably that the cement is readily carried to all parts of the root. Where the natural crown is still in place, it might not always be possible to thoroughly fill the canal with such a plastic material. Some claim that oxyphosphate lacks the virtues of oxychloride, but the statement, however authoritatively asserted, would be hard to prove. There is one objection to either which is important. It is difficult to remove them, should it be desirable to empty the canal.

Gutta-Percha.—This material, in some form, is used by the majority of dentists, and rightfully so. The usual custom is to roll the white variety into cones, which are slightly warmed and pressed into the canal. Chloro-percha is most useful when not relied upon alone. Forced into a canal and followed with a cone, it renders the operation of filling oftentimes more easy and more perfect.

Other materials have from time to time been advocated, but none so generally adopted as the ones I have named.

The first condition in which a tooth requires root-filling is where the pulp is freshly removed. Here several phases obtain.

The pulp may be removed fully, coming away so that the operator is thoroughly satisfied that there is none left behind. Here is the ideal place for immediate root-filling. Where the canal is accessible, there is no need for a reamer ; but if it be attenuated or irregular, the reamer is to be cautiously used as has been described. The dam being on, I would use a hot-air syringe, drying the cavity thoroughly, continuing the blast until the patient complains of the heated condition of the tooth. Then I insert a cone of Gilbert's temporary stopping, which I have previously dipped into chloro-percha, and force this up into the canal. Where I am sure that the canal is hot, I do not heat the cone. In a stiff condition it is more readily forced into place, and the heated walls soften it so that it is easily packed to the end. This, however, should be done with the utmost care, and the slightest expression of pain from the patient, such as wrinkles between the eyebrows, is a sign to stop. It is a danger signal which, when obeyed, may prevent pressing the material through the apex.

It is to be observed that at the very outset I advocate a material not mentioned in the list upon which I made comments. Still, temporary stopping is a form of gutta-percha, being only that material mixed with wax. I prefer it to the gutta-percha alone, for the reason that it softens at a much lower temperature, and remains so longer. Where I have succeeded in getting the canal-walls properly heated I have often inserted my cone cold, and in a moment found it nearly as plastic as soft wax. It is also less liable to change its shape either by shrinkage or expansion, chloro-percha having been often accused of the one and gutta-percha of the other.

Where the pulp is taken out bit by bit, the reamer is more necessary, as where it can be used it will remove the odontoblasts adherent to the walls.

Where the removal of the pulp is followed by hemorrhage, immediate root-filling is contraindicated, save where the hemorrhage is controllable, as it often is by the packing of cotton dipped in tannin and glycerin or other styptic into the canal, and leaving the dressing there for ten minutes. If the hemorrhage persists, a similar dressing may be left in for twenty-four hours. It should be removed with great care. Drawing it out slowly until it is loosened, and then syringing thoroughly with warm water, will often disengage the cotton from the clot which has formed, and which, if suddenly torn away, will start a new hemorrhage. The clot itself may be removed with a little perseverance, using cotton twisted on a broach and dipped in warm water. This softens it and cleanses the canal. One who has not done this, resting satisfied with simply removing the dressing, will be astonished to see how many twists of cotton will be stained with blood before one can be removed clean.

Again, we have a condition where there will persist a sensitiveness at or near the apex. This may be present, despite the best treatment, for several days, on any one of which the insertion of a root-filling would bring the patient hurrying back to the office within a few hours in great pain. This condition is somewhat mysterious, and the most plausible explanation is as follows : It is not likely that a pulp may be destroyed with arsenic, and when removed leave a portion of the extremity in place, and this be alive. Such a remnant, however, may be left, which would then be connected with the tissues beyond. Where arsenic has been used liberally, or where the patient is easily affected by the drug, it is not difficult to understand that it may have exerted an influence beyond the apex, so that the pericementum has become irritated therefrom. Any pressure against this remnant of pulp, whether by instrument, filling-material, or a column of air, would be conveyed to the tissues beyond the apex, which would respond. Any attempt to fill immediately is an error. The canal should be dressed lightly with an anodyne, and the cavity sealed up, and left for a week or longer. This allows the irritation time to subside, and then filling may be attempted with little danger. Nevertheless, I deem it wiser to be on the safe side here, and have adopted a method which I think is peculiarly my own, but which is nevertheless thoroughly satisfactory.

Take floss silk and wax it thoroughly, after which dip it into chloro-percha and cut it into pieces about an inch long. These, when dry, give us gutta-percha cones which have a silk through them. *They are readily packed into a canal, and the end being allowed to extend beyond the orifice of the canal, is readily grasped, in case of need, with a pair of tweezers, whereupon the whole root-filling is easily withdrawn.* Where no trouble ensues, the root-filling of this kind may safely be left in place, being quite dissimilar from cotton, as the silk fiber is thoroughly incased in gutta-percha.

The next condition is where the pulp has died without the aid of a dentist, and the patient presents with a pericementitis. The treatment is to abort the disease if possible, or in case this fails, hurry it on to the suppurative stage. Where the pericementitis seems to subside without suppurating, I prefer the silk-and-gutta-percha cones, because of their easy removal.

Next we have teeth presenting with abscesses. If the abscess is acute, I think it wiser to cure it before filling the root, except where the tooth is single-rooted, and it is decided to amputate the apex and remove it with the abscess attached. Then it is best to fill the root at once, so that the amputation will not leave an empty canal with a gaping opening at the end. These teeth should be filled with the temporary stopping packed solidly into the root. The amputation is accomplished thus: Select a spear-drill, and with it pass through gum, process, and root, along its central axis as high up as is decided for the amputation. Follow this hole with a sharp fissure-bur, with which, by laterally cutting first one way and then the opposite, the end of the root is easily severed. Where there has been much destruction of the process, its removal will not be difficult. Where, however, this has not occurred, as, for example, in a cuspid where there is a long root and dense alveolus, it may be so difficult to do this that it would be wiser to anesthetize the patient. Then the labial plate of the alveolus may be removed with burs and the amputated root-end extracted.

In chronic abscesses I scarcely ever treat other than surgically, so that immediate root-filling is more permissible here. Still, I prefer the silk and gutta-percha in all cases where there is the least chance that some day this must be removed.

It may be argued that I should not advocate seemingly temporary methods; but while it is true that we should hope to make our work as permanent as possible, in the matter of root-filling, too positive permanence is a detriment rather than an advantage. It never can be certainly asserted of any tooth that its roots will never need to be unfilled. If in no other way, the natural crown may continue to decay till it is lost, when a crowning process may make it imperative to empty the canal or canals. Where they are found filled with a very resistant material, there will always be some difficulty experienced. Again, *I have seen teeth lost, where pericementitis had set in, which could not be adequately treated because the root-canals were so filled that they could not be emptied, the teeth being too sore to the touch to make the necessary drilling possible.* When such a case presents the dentist will quickly say to himself, "I wish this root were filled with temporary stopping, which I could remove with a heated instrument." Of course the other method, using the silk-and-gutta-percha cone, is satisfactory.

Either of these methods require that at least a slight layer of oxyphosphate or oxychloride should cover them before gold is packed upon them. This will be unnecessary where amalgam is to be depended upon.

In concluding this work, I have only to state that in describing the methods that I have successfully used I do so with no special desire to impress my readers with the idea that they are my methods. It seems to me immaterial who originates a method. The main thing is that it be useful. I have learned nearly all that I know from others, and from experience. If I can now teach any one, it will be in some degree a repayment of the debt.

INDEX.

ABRADED teeth, etiology of, 132.
  filling of, 132.
  use of screws in, 100.
Abscessed teeth, treatment of, 195.
Alloys, mixing of, 56, 61.
Amalgam fillings, finishing of, 175.
  use of matrix in, 47, 57.
  where to use them, 58.
Amalgams, combination of with other materials, 60.
  in contour fillings, 95, 147.
  method of use, 56.
  rapid setting of, 58.
  relative values of, 55.
  shrinkage of, 55.
Ambidexterity, advantage of to the dentist, 82.
Anchorage, difficulty of with sensitive dentine, 125.
Approximal cavities, extension of, 15.
  preparation of, 5. 104, 121.
  rule for filling, 106.
Approximal trimmer, use of in finishing fillings, 174.
Arrested decay, 53.
Arsenical dressings, covering of, 171.
Arsenical poisoning, treatment of, 171.
Artistic work, 128.
Automatic mallet, use of, 81.

BAND, use of in filling-operations, 149, 168.
  in restoration of crowns, 103.
Beveling enamel margins, 26, 122, 163.
Bibulous paper, control of moisture by, 43.
  protection of gum-tissue by, 60.
Bicuspid crown, relation of to the dental arch, 119.
Bicuspids, filling of, 90, 101, 118, 121, 137, 161.
  roots of, 187.
Bite, opening of, 134.
Bonwill mechanical mallet, 82.
Bottomless cavities, 164.

Bridge-piece, retention of by filling, 150.
Buccal cavities, filling of, 166.
Burnisher, use of in finishing fillings, 175.
  on sensitive teeth, 153.
Burs, caution in use of, 23.

CARIES, susceptibility of human teeth to, 83.
Cavities, classification of, 104, 151.
  general principles involved in preparation of, 2.
  intentional extension of, 9, 13, 105, 115, 120, 125, 127.
  methods of keeping dry, 27.
  special principles involved in preparation of, 104, 126, 151.
Cavity borders, formation of, 18.
Cement fillings, advantages of, 53.
  disadvantages of, 62.
Cements as root-fillings, 192.
  modes of mixing, 53, 54.
  where to use, 55, 76, 97, 99, 147.
Cementum, sensitiveness of, 152.
Central incisor, abrasion in, 132.
  approximal cavities in, 88, 105.
  contour filling of, 94, 97.
  restoration of cutting-edge, 14.
  root of, 177.
Children's permanent teeth, filling of, 99.
Chisel, use of in preparation of cavities, 140.
Chloride of zinc as a germicide, 52.
Chloro-percha as a root-filling, 193.
  cavity lining with, 51.
  control of moisture by, 43, 175.
Clamps, application of, 37.
Cocaine, use of in applying ligatures, 31.
Cohesive gold, manipulation of, 64.
Compound cavities, 122.
Contour fillings, burnishing of, 74.
  manipulation of, 85, 92.
  use of amalgam in, 57.
Copper amalgam, therapeutic value of, 62.

Corners, fracture of by hand-pressure, 80.
  restoration of, 20, 109, 111, 112, 113, 121, 125, 129.
Cotton as a root-filling, 191.
  as a temporary filling, 172.
  control of moisture by, 43.
  use of in festoon cavities, 30, 160.
Crown, misuse of term, 104.
  restoration of, 101, 102, 103.
Crown cavities, choice of materials for filling, 59.
  enlargement of, 145.
  preparation of, 8, 137.
Crystal gold, 65, 129.
  methods of using, 66.
Curved roots, use of drill in, 189.
Cuspids, contour filling of, 117.
  festoon cavity in, 161.
  incisive edge of, 135.
  microscopical section of, 152.
  roots of, 187.
Cusps, restoration of with amalgam, 147.
Cutting-edges, reproduction of, 91, 94, 111.

Decay, removal of, 2, 136.
Dentine, preservation of in formation of cavities, 131.
  sensitiveness of, 76, 152.
Disk-carrier, best form of, 174.
Disks, use of in finishing fillings, 25, 173, 175.
Distal cavities in cuspids, 118.
"Double teeth," 133.
Drill, use of in root-canals, 188.

Economy, true and false, 68, 70.
Enamel, cleavage of, 21.
  contact of gold with, 131.
  effect of green-stain on, 158.
  sensitiveness of, 152.
Enamel margins, 24.
Erosion, caution in diagnosis, 154.
  distinction of from abrasion, 132.
  etiology of, 155.
  treatment of, 157.
Evans clamp, proper use of, 38.
Evans root-drier, 188.

Festoon cavities, preparation of, 8, 160.
Figure-of-8 ligature, 34.
File, use of, 84, 174.
File-marks, avoidance of, 60.
Filling-materials, varieties of, 47.
Finishing of fillings, 173.
Fissure cavities, preparation of, 10.
Flat fillings, 89, 92.
Foot-plugger, use of, 67, 73, 75, 93, 123.

Fracture of amalgam fillings, 96.
Fractured teeth, filling of, 35, 113, 130, 134.
Frosted gold foil, 67.

Gates-Glidden drill, 185.
Germicides, combination of with oxyphosphates, 54.
Gilbert's temporary stopping, 49, 193.
Gingival border, failure of fillings at, 39, 59, 118, 122.
Glass as a filling-material, 49.
Gold as a filling-material, 63, 120.
  as a root-filling, 191.
  combination of with oxyphosphate, 75.
  contour fillings, 92.
  how to condense, 79.
  incorporation of with amalgam, 57, 60.
  unsuitable cases for, 165.
Gold and iridium, 79.
Gold and platinum, 77.
Gold and tin, 78.
Gold fillings, finishing of, 173.
  for crown cavities, 59.
  leakage of, 40, 60.
  use of matrix in, 46.
Gravitation, effect of on fillings, 108, 112.
Green-stain, 157, 168.
Grinding-surfaces, reproduction of, 90.
Grooved incisors, treatment of, 14.
Grooved teeth, illustrations of, 156.
Gum-recession, root-exposure from, 170.
  tooth-sensitiveness from, 151.
Gum-tissue, arrangement of in man, 84.
Gutta-percha as a filling-material, 49, 172.
  as a root-filling, 192.
  choice of colors, 50, 171.
  cones for root-filling, 194.
  finishing of, 175.

Hand-mallet, advantages of, 81, 161.
Hand-pressure, danger of fracture by, 111.
  in gold fillings, 79, 111.
Hart, J. I., on sensitive dentine, 153.
Heavy foil, manipulation of, 74, 93.
  uses of, 71, 122, 161.
Heitzmann, C., microscopical section of cuspid, 152.
Hemorrhage, control of, 193.
Hollow fillings, 72, 92.
Horse-shoe grooves, 141, 164.
How cervix clamp, 39.
How screws, use of, 100.
Hypersensitive teeth, gold fillings in, 76.

Incisive edges, cavities in, 126, 134.
  contouring of, 131, 135.

INDEX.

Incisors, filling of, 14, 86, 94, 97, 100, 101, 105, 133.
   root-filling of, 177.
Instruments, use of, 83, 161.
Interzonal layer, sensitiveness of, 153.
Iridium, combination of with gold, 79.

KNOTS, methods of tying, 32.

LABIAL cavities, filling of, 73, 106, 110, 123, 162.
Lateral incisor, proneness of to abscess, 116.
   root of, 178.
Lead as a filling-material, 48.
   as a root-filling, 191.
Leakage of fillings, 39, 60.
Ligatures, different forms of, 31.
Lingual cavities, filling of, 165.
Loose teeth, filling of, 33.

MALLET, choice of, 81, 161.
Mastication, effect of on the teeth, 86, 91.
   fracture of fillings by, 96.
Matrices, uses and dangers of, 45, 90, 119, 173.
Milk, production of green-stain by, 158.
Mixing-slab for oxyphosphates, 53.
Moisture, devices for controlling, 43.
   in contour work, 96.
   recurrence of, 40.
Molar crown, relation of to the dental arch, 119.
Molars, enlargement of cavities in, 11.
   filling of, 91, 99, 102, 124.
   roots of, 189.
Mouth, examination of, 138, 143.
Mouth-mirror, use of, 108, 110, 111.

NAPKINS, use of, 42.
Nitrate of silver, use of on sensitive teeth, 154.
Non-cohesive gold, use of, 64.

OXYCHLORIDE of zinc as a filling-material, 51.
Oxyphosphate of zinc, combination of with amalgam, 61.
   combination of with gold, 75, 87, 147, 165.
   finishing of, 175.
   in contour work, 97.
   in crown-setting, 54.
   in festoon cavities, 160.
   manipulation of, 53.

PALATAL cavities, filling of, 164.

Palato-approximal cavities, filling of, 110, 123, 166.
Pellets for small cavities, 106, 167.
Pettit reamer, 189.
Phosphate, combination of with amalgam, 61.
Piano-wire, use of in root-filling, 187.
Pin-head fillings, improper use of, 11.
Pipe-clay disks for control of moisture, 43.
Plastic gold, advantages of, 65.
   combination of with amalgam, 61.
Platinum, combination of with gold, 77.
Pluggers, choice of, 82.
Polishing, best methods of, 25, 173, 175.
Porcelain fillings, 49.
Porcelain inlays, 49.
Power mallet, unsuitable places for, 130, 161.
Pressure, influence of on gold, 80.
   line of, 97, 142.
Probational fillings, 170.
Pulp, retention of dentine over, 136.
   size of determined by age, 124.
Pulp-capping with gutta-percha, 51.
Pyorrhea alveolaris resulting from wedging, 85.

REAMER, use of in root-canals, 187.
Retaining-points, 112, 116, 123, 142, 167.
Retentive shaping of cavities, 4, 120, 127.
Robinson's felt, 48.
Root-canals, preparation of, 183.
Root-filling, difficulties of, 176.
Roots, exposure of from gum-recession, 170.
Rose bur, use of in preparation of cavities. 105, 115, 121, 126, 127, 130, 133.
Rubber cup, use of in finishing fillings, 174.
Rubber-dam, placing of, 28, 41.
   repair of, 42.
Rubber tubing, use of in festoon cavities, 30, 160.
Rubber wedge, use of, 44.

SANDARAC varnish, objections to, 171.
Saucer-shaped cavities, 6, 72, 125.
Screw-mandrel, advantages of, 174.
Screws, use of, 99, 113, 114, 123, 134.
Secondary dentine, formation of, 52.
"Self-cleansing" surfaces, 16.
Separators, abuse of, 43, 119.
Silk and gutta-percha cones, 194.
Sixth-year molar, filling of, 99.
Soaping disks, 42.
Space, necessity for in filling-operations, 119, 120.

Sulci, reproduction of, 90.
Surface cavities, 155.
　rarity of in incisors, 159.

TAPE separators, 44.
Teeth, change in position of, 91.
　decalcification of by green-stain, 169.
　occlusion of, 84.
　separation of, 84.
　shortening of by abrasion, 133.
　union of by a single filling, 150.
Temporary fillings, 170.
Temporary stopping, use of, 49, 171.
Tin as a filling-material, 48, 170.
　combination of with gold, 78.
Tooth-brush, effect of on tooth-structure, 153.

Tooth-neck, sensitiveness of, 151.

UNDERCUTS, filling of, 69.
　use of in preparation of cavities, 6, 8, 94, 105, 112, 117, 142.

VITREOUS fillings, 49.
V-space, injurious results from, 85.

WATTS'S crystal gold, 65.
"Weaving" method in ligatures, 36.
Wedges, use of, 43.
Wheel-bur, use of in extension of cavities, 126.
Wisdom-teeth, roots of, 182.
Wood as a root-filling, 191.
Wooden wedges, 45.

www.ingramcontent.com/pod-product-compliance
Lightning Source LLC
Chambersburg PA
CBHW020903230426
43666CB00008B/1299